COOKED

COOKED

FROM THE STREETS TO THE STOVE,
FROM COCAINE TO FOIE GRAS

JEFF HENDERSON

WILLIAM MORROW
An Imprint of HarperCollins*Publishers*

Grateful acknowledgment is made to the following for the use of the interior photographs: insert pages 1–15, courtesy of the author's collection; page 16, courtesy of Bellagio/MGM Mirage.

HarperCollins books may be purchased for educational, business, or sales promotional use. For information please write: Special Markets Department, HarperCollins Publishers, 10 East 53rd Street, New York, NY 10022.

FIRST EDITION

Designed by Judith Stagnitto Abbate/Abbate Design

Library of Congress Cataloging-in-Publication Data

Henderson, Jeff.
    Cooked : from the streets to the stove, from cocaine to foie gras / by Jeff Henderson.—1st ed.
        p. cm.
    ISBN: 978-0-06-115390-7
    ISBN-10: 0-06-115390-7
    1. Henderson, Jeff. 2. Drug dealers—United States—Biography. 3. Cooks—United States—Biography. 4. Ex-convicts—United States—Biography. I. Title.

HV5805.H37 A3 2007
364.1'77092—dc22
[B]
2006052145

07 08 09 10 11 JTC/RRD 10 9 8 7 6 5 4 3 2 1

This book is dedicated to
my grandmother, my family, and my community

CONTENTS

ONE SIX COURSES IN SIXTY MINUTES 1

TWO POCKET CHANGE 13

THREE T-ROW 27

FOUR MY FIRST ROLEX 39

FIVE CAUGHT UP 63

SIX SOUL ON ICE 77

SEVEN TERMINAL ISLAND 101

EIGHT HARD HEAD NO MORE 117

NINE KITCHEN HUSTLE 131

TEN DAYLIGHT 149

ELEVEN MEAL TICKET 159

TWELVE THE GREAT GADSBY 179

THIRTEEN WORKING THE PASS 201

FOURTEEN SABOTAGE 219

FIFTEEN CHEF OF THE YEAR 239

AFTERWORD GOD BLESS THE DEAD 263

ACKNOWLEDGMENTS 273

A NOTE FROM CHEF JEFF

This is a work of nonfiction. It is my story as I remember it. All of the events and experiences detailed are true and have been put down as I recall them. Some names, identities, and circumstances have been changed in order to protect the integrity and/or privacy of the various individuals involved.

Conversations throughout this work have been recorded as I remember them, but they have not been written to represent word-for-word documentation; rather, I've retold them in a way that evokes the real feeling and meaning of what was said, in keeping with the true essence of the mood and spirit of the event.

COOKED

ONE

SIX COURSES IN SIXTY MINUTES

By the time I showed up in Las Vegas, I'd been looking for work for more than a month. I had busted my ass in the five years since my prison release, rising from dishwasher at a small restaurant to sous-chef at one of the most prestigious kitchens in L.A. I was on track toward running my own restaurant when a political kitchen battle suddenly left me begging for someone to give me a chance to start over. I hadn't been jobless this long since I'd left prison, and my prospects of landing a position hadn't been so bleak since then, either.

Every potential employer I met with seemed only interested in the fact that I was a convicted felon. They didn't care that I'd proven myself in some of L.A.'s best kitchens or that I really could cook. They definitely didn't care that I had a wife and two young children to support, and that I'd spent the last of our savings on a one-way ticket to the desert hoping to restart my career. A week into my search, every hotel on the strip had turned me down.

When I visited these properties, most of the people I interviewed with liked me. My cooking resume was impeccable, five stars across the board, but their enthusiasm had a way of drying up as soon as I told them I had spent time in federal prison for drug trafficking.

On the outside, I was what was acceptable for a black man in corporate America: clean shaven, earring hole covered up; I even toned down my walk so that I wouldn't swagger and come off as ghetto during interviews—I've got a pretty good stroll.

Still, it always came down to me being a felon. Everywhere I went, they gave me this smoke-and-mirrors bullshit, telling me, "We'll call you when we're ready." At the Paris Hotel, they were introducing me to my staff before I told them about my criminal record. Then they told me to take a walk.

With potential employers, I always explained about my past: I was young, I made some mistakes, and I spent years regretting those mistakes. My criminal past was so far behind me that I regularly lectured schoolkids about how crack had been destroying our community since back when I was just a schoolkid myself. None of these execs were having it—like I was the first ex-con who ever looked for work on the strip.

By the time I showed up at Caesars Palace, I was desperate.

Caesars was a place I knew well because I used to roll there when I was a dealer. Back in the day, no one knew how to cater to high rollers like Caesars. Me and my boys used to come up from California for all the prizefights with Louis Vuitton bags full of cash. We gambled thirty Gs at a whop. And Caesars management? They loved our asses. We flew in and a limousine driver was holding up a sign at the airport for the "Henderson Group."

But "back in the day" was fourteen years back already, and I didn't have any Louis Vuitton bags. I sure didn't have one full of cash.

The night before my Caesars interview, I snooped all over the hotel to put my game plan together. If I saw some cooks walk into the casino, I would roll up on them.

"Hey, how you doing?" I'd say. "My name's Jeff Henderson. Can I talk to you for a second? I'm thinking of moving up here. What's it like? What's the chef like?"

It was a reconnaissance mission. Since I'd have to prepare a tasting meal for the executive chef, I planned to base it on the foods he liked. I wanted to make my mark by showing up for the interview with the full menu in my briefcase. So when he says, "Hey, this is nice," he doesn't know that I've already been on his property eating his food. The cooks tell me he likes Italian, so I go to the Caesars Italian restaurant, Terrazza, and have the Veal Milanese. I even chatted up some of the hostesses to get a feel for the hotel politics.

By the time I walked into the man's office, I was comfortable, confident. It was a huge room decorated from one end to the other with Roman-style artifacts, the walls covered with pictures of prizefighters. The man behind the desk was a smooth middle-aged Italian from New York with black hair slicked straight back.

And here I was, this black motherfucker in a $150 Brigard chef's coat made of Egyptian cotton. I went right into my hard sell, telling him that I was ready to go to work on the spot. I told him straight up: "Look, Chef, I've done some time. I learned to run a kitchen in prison. But my resume speaks for itself."

I think he liked my aggressive approach. In Vegas, like in prison, you have to be tough to run a kitchen. If the cooks sense any sign of weakness, they'll run you over, tell you how to do your fucking job.

"Mr. Henderson," he said. "Did you ever kill anyone?"

"No, sir."

"All right," he said. "I want you to cook me dinner on Friday. Write up a menu."

I opened my briefcase, showed him the menu I'd already typed up and brought along with me, and told him that instead of giving me the usual ninety-day probation period, just to give me a month.

"That won't be necessary," he told me. "Just cook me a tasting dinner for six."

That tasting dinner would be a tryout for the food and beverage executives. Six courses in sixty minutes would decide my fate and the fate of my family. It would be the most important meal I ever cooked.

I remember I had my game face on, moving up and down the line in that sprawling kitchen like a general on the battlefield, flames roaring from my stove.

After I served them an amuse bouche that came out perfect—a beautiful pan-seared U-10 diver scallop with a white truffle creamed corn sauce—my confidence was high. They were impressed. My timing was on point as I was plating the first course, a microgreen and roasted-pear salad with gorgonzola. I knew I had them on the ropes as I plated the next course, Hudson Valley foie gras served with warm minted pineapple. That's when I realized my fucking foie gras had been sitting out for about thirty minutes and started to oversoften.

Two things you need to know about foie gras: It is incredibly expensive and absolutely unforgiving in its delicacy. Foie gras has the consistency of butter and can turn into a useless mush if left out in a hot kitchen. With a great piece of steak or even lobster, you can screw up and there are ways to cover it up so that no one will notice. That's not the case with foie gras. Just like when you're cooking cocaine, one miscalculation of heat can destroy your product.

With crack, if you don't babysit the pot and micromanage the process, a third of your yield can disintegrate. And with a kilo of cocaine selling for $14,500 wholesale, that is not an acceptable loss.

Back in the dry spell of '86, I couldn't afford to lose so much as a gram. I was twenty-one, and I had a lot of clients in San Diego counting on me to get them some work—when I say "work," I mean make a buy. But all my connections in L.A. had dried up. Even my most reliable source, this pair of rich twin brothers, was sold out. But I knew some guys who owned a car dealership outside of Beverly Hills and I thought they might be able to hook me up.

All the black dealers would buy their high-end cars from those guys because they'd take care of the paperwork. Normally, we couldn't buy cars from a dealership because legit dealers—*car dealers*—had to report any cash transaction over $10,000 to the IRS. These guys, though, they'd hook it up to look like you were making payments.

I had my sister drive me out to see them. Skinny as a cigarette, we called her Cali Slims. She was my most trusted confidante, so I always took care of my sis. But I didn't want to give her money straight out. Instead, I paid her to drive for me— $1,000 from San Diego to L.A. and back.

When I first approached the boys, they didn't want to deal with me. They still wanted to play it off like they didn't know what their cash customers did for a living. So I'm, like, "Listen. I need some work." I had to talk in circles for a while but finally they hooked me up with this Mexican dude by the name of Cholo.

What was happening was that all the Colombians kept getting busted, so they started to use Mexicans to run their blow into L.A. Now Cholo was real Rico Suave, but he wasn't intimidating. His crew, though, looked like some tough motherfuckers.

Cholo wouldn't fuck with you unless you were buying ten birds or more—we always called kilos "birds," or "chickens," or "them things." Like I said, you had to be buying at least ten of them, and someone had to vouch for you. Cholo would have to talk to someone who told him I was cool, that I had shit sewn up and could have a lot of work. So I went back to L.A. to wait while the shady car dealers vouched for me.

A few days later we had a meeting set up at a Denny's restaurant. I was getting ten birds at $14.5 Gs apiece. The money was in my Louis Vuitton: $145,000 in thousand-dollar packs of fifty- and hundred-dollar bills wrapped in rubber bands. My clients were small dealers who paid me in everything from singles to twenties, but you don't fuck with small bills when you're buying bulk.

I knew this girl Paula who ran a check-cashing place. On the first and fifteenth of every month everyone would cash their welfare checks. The night before, I'd go to her crib and trade a hundred grand in singles, tens, and twenties for the clean fifties and hundreds that she had just gotten off the money truck. I'd kick her a little taste and my money would be clean. I'd have the crisp bills that were easier to make deals with, and her customers would get the money that my people had gotten off the streets. Of course, within a week, many of the customers of the check-cashing place would bring their county money back to our crack houses and we'd take that money right back to the white man, buying all the flashy shit a hustler had to have. And, if we're caught, the DEA takes all *our* shit and sells it back at auction. It's a fucking game.

Anyway, my sister and I are in the parking lot of this Denny's waiting on Cholo when this Mexican dude I'd never seen before rolls up beside us in a 300 ZX.

He says, "You looking for Cholo?"

"Yeah," I tell him, and I'm already starting to get nervous.

"Come with me," he says. I looked at my sister like, What the fuck? At this point, I'm scared, but I don't want to bitch up, so I just get in his car. I was hoping my sister would take down his license plate.

For the next half hour this dude doesn't say shit, just drives me around, zigzagging through the hills so that I don't know how to get where we're going—or back. Finally he pulls over next to this nice middle-class-looking residence and borrows my cell phone, this big-ass Motorola NEC that was heavy as fuck. He speaks some Spanish, hands me back my phone, the garage opens up, and we roll in.

There's Cholo.

"You want ten?" Cholo says.

"Yeah."

The weird shit is, he doesn't pat me down for weapons or any-thing, just leads me into a family room, where I see his arsenal: a whole motherfucking *bunch* of Mexicans. There must've been ten or twelve of them and I damn near pissed myself. I was a young, skinny, light-skinned youngster and it would've been easy for them to take me out or fuck me up.

Cholo points to a chair, goes "Sit down," and right about then I had to because my knees were about to start buckling. They count my money and Cholo opens up a closet full of kilos. There were at least two hundred birds up in there. You do the math.

To my relief, Cholo brings out ten kilos—Peruvian pearl white, each brick with "Rolex" written across its wrapping in black marker. Rolex was good dope back then. They put the keys in a Foot Locker bag and the dude who'd driven me there gets up, say-ing, "Let's go."

"All right, Cholo," I said. "I'll get back at you."

"You gonna keep buying from me, right?"

"Fuck yeah," I told him. "I'm gonna call you."

I was never coming back. I wanted to get back with my twins soon as shit—but first I wanted to get the fuck out of Cholo's place.

Back in San Diego, I dropped my sister at her crib and switched into my work vehicle, a white '85 Chevy Celebrity with burgundy interior—a clean-cut car like your grandmother would roll in. I paged my boy Michael. He was going to buy four of my birds and I was going to teach him how to cook it.

Michael comes over, we drive down to the Safeway and I buy all their baking soda, maybe twenty big boxes, and most of their sixteen-ounce GladBags.

Then it was on to Kmart, where I bought three sets of Corning Ware Pyrex pots, because I wanted to cook three kilos at a time. Whenever I did a big cook, I'd buy all new equipment and throw it away when I was done. At the 7-11, we picked up fifteen bags of ice and headed over to the Motel Six in Spring Valley. I already had my triple-beam scale in the trunk with the drugs.

I always cooked my dope in Spring Valley because for some reason all the hotels down there had four-burner stoves right across from the bathtub, where I'd cool down the dope. We brought all our stuff into the motel and chilled for a minute while I schooled Michael on the art of cooking dope. He was lucky that he had someone to teach him for free. I learned by looking over the shoulders of the Twins and L.A. Will, and then experimenting on my own. Other dealers were selling the recipe for thousands of dollars.

Once we'd finished the cook, we got on to weighing and bagging. It took a good two hours to bag up all those kilos. Then I started calling all my clients to let them know I'm in pocket. My cell phone and pager started blowing up and I was making drops all over the city. Everyone was calling in their orders.

When San Diego is dry and I'm the only one with an L.A. connection, I can charge anything I want, because I'm the only game in town.

My last drop was a guy by the name of Six-Six. He drove a candy apple red '66 Chevy Impala and he lived on the East Side. I was bringing him a kilo of hard.

Six-Six had a crack house on Fifty-fourth and Imperial, which was the crack Mecca of southeast San Diego in the mideighties. People came from all over to dump their stuff on that particular neighborhood—Bloods, Crips, everyone. And everyone showed up there to score rock, from gangstas to suburban kids, hookers to lawyers.

My woman Carmen and me pulled up in the alley behind Six-Six's crack house, which was a whole strip of Section 8 apartments between two other crack houses. I sold a kilo of hard for $16,500 to customers who paid me up front, but Six-Six was buying on consignment so he'd end up paying an extra $2,000 in interest.

Carmen and me go up to his place, Six-Six puts on some music, I give him his key, and he gives me the $18,500 for the work I'd given him with another kilo. So there we are kicking it, listening to music and counting money. I'm just chilling for the first time in a week. I've made all my money back, plus the return. In a few days I'll have to go re-up again, start the whole cycle over, but for now I'm going to relax.

That's when a lot of pounding and noise outside started.

"Shit!" Six-Six is yelling. "What the fuck?"

I peek through the curtains and see motherfuckers in blue police task force jackets and 9 mm pistols busting in on the apartment across the way. I don't even have a chance to shout "It's a raid!" before I turn around and see Six-Six booking for the bathroom.

Six-Six was crouched over the toilet, tearing apart his key— the one he still hadn't paid for—and dumping every gram into the bowl. Carmen and me, our asses were out the side window.

I'm known to the police, so we split up.

Carmen takes the car and I take the money, running for the city bus. The sweat is streaming down my face and my heart's beating like fuck while I ride the bus toward home, wondering what happened to Six-Six.

It turned out they broke down his front door, but he was already out a side window and they never found any evidence of that kilo he flushed.

Even taking an $18,500 loss on that last key, I'd still come out too far ahead to bitch about it. I'd bought ten kilos of coke for $145,000, blown them up to fifteen kilos of crack, and sold fourteen of those at $16,500 each for a total of $231,000. In less than a week I'd made a profit of $86,000. And I was still a free man. That's what I called an acceptable write-off. But, in that business, nobody's luck holds out for long.

Staring down at the melting liver, I knew that its thin outer membrane might burn away the second it touched my hot sauté pan, its contents oozing out as a shapeless mess to be thrown in the garbage. Even if the membrane somehow managed to hold, the foie was in such a vulnerable state that the whole thing would most likely overcook in seconds. I wouldn't be able to serve it. And it would take me too long to get a replacement course out to them. They'd know for sure that something wasn't right.

I was starting to lose my shit. I had to self-talk, change the presentation, disregard the course. I had to do something.

I paced the hot line, watching the execs and the other chefs eyeing me over their plates, thinking about how close I was to blowing my shot, about my wife and my two little kids back in L.A., waiting to hear if I'd found a job or not. I could almost hear the words coming out of my mouth, telling my wife that it was over. After all my years of hard work, all we'd gone through together, I'd failed. My dream was dead. I just couldn't do it; I couldn't skip the foie gras. It was the centerpiece of the whole meal. I had to go for it.

Everything was riding on this. I'd get one chance. The pan would have to be screaming hot. I'd have to slide the foie medallions in carefully and be ready to gently turn them almost immediately. One shot, my last shot.

Spooning a cube of butter into a sauté pan, I added minted pineapple, brown sugar, and molasses, which I'd use to glaze my foie gras. In a cast-iron skillet I quickly sautéed some pineapple and set it aside. Then, I got that skillet blazing hot and drizzled in an oil blend (60/40 vegetable oil to olive oil). Just before the oil started smoking, I carefully placed my foie into the pan. As I laid each piece down I immediately, gently, flipped each one. I had to seal each piece, so it wouldn't become mush. Seconds later, it was done. I glazed all of the pieces and plated them up.

The dish knocked everyone out. Even with another five courses to go, I had everything under control. I'd come a long-ass way to cook dinner for these corporate boys.

The rest of the tasting was banging. After the foie gras came my fish course. I did oven-roasted striped sea bass with savoy cabbage and fingerling potatoes infused with chive olive oil. For my meat dish, I did filet mignon that was plated with caramelized onion on top of the meat and celeriac whipped potatoes (a potato puree blended with celery root puree).

The first dessert was strawberry and doughnut-peach soup. On the second dessert I gave them deep-fried banana fritters with warm caramel sauce and a vanilla bean milk shake.

They had probably never seen anything like me: a black man who could really cook. I think they were expecting some hack, but I gave them a meal they'd remember. They were blown away. Even the other cooks in the kitchen were impressed. As I packed up my knife kit, the chef came up to me and said, "Jeff, how long will it take you to report for work?"

I told him to give me a week to wrap things up in L.A.

When I left Caesars that night I was relieved. But that sense of ease and accomplishment started turning into anxiety as soon I started to think about my situation. I had pulled back from the edge, used my skills to resurrect my career. Caesars was offering me a great new beginning. But the pressure was still there: One misstep could quickly mark the end of everything I had been working so hard to achieve.

TWO

~~~~

POCKET CHANGE

My father left when I was one or two years old. It was just my mother, my sister, Cali Slim, and me living in South Central. I only saw my dad on holidays and the occasional weekend, but since most of Moms's time was spent working as a home care provider for a quadriplegic woman in the Valley, my father's parents played an instrumental role in my upbringing.

My grandfather owned his own janitorial business, and he started taking me on the job with him when I was about five years old. He taught me to work hard and he taught me the value of a dollar.

And he taught me to steal.

My grandparents had migrated to California in the late fifties looking for better opportunities. My grandmother was a Creole woman who almost passed as white and so got one of the better jobs cleaning houses up in Beverly Hills. My grandfather was a dark black man originally from Mobile, Alabama. He was a stern, strict man who was a hard worker, but rumor was he always kept a private nightlife going. In New Orleans, he had worked as a long-shoreman on the Mississippi River. When he came to California, he was frustrated because he had to deal with white people on a regular basis. Back in New Orleans, everyone he worked with was black and he only encountered white people when he picked up his paycheck.

I remember how he used to load his ladder and all of his cleaning equipment into the back of his gold '65 Chevy and hustle all over Los Angeles getting commercial gigs, so that he could clean their businesses, change fluorescent lights, wash their windows.

When I was around six or seven, I started helping him out. We would clean all these Baskin-Robbins, Zieler and Zieler's clothing stores, and different bakeries up in the Wilshire district, a prominent Jewish community we called Jewtown. But my grandfather's main business was with Laundromats in L.A., cleaning them and installing new machines. It seemed to me like he had the keys to every Laundromat in the county. I vividly remember that whenever we were getting ready to leave he'd always tell me to wait by the door. I'd peek around the corner and see him taking quarters out of the machines.

I was, like, "Damn!" Part of me knew it was wrong, but then again, I really didn't understand what it was about. When we got back to my granddaddy's house I'd watch him take the quarters and put them into paper rolls, so I knew there was all this money right on the living room mantel that didn't really belong to him. Of course, that's not how Granddaddy saw it, and I discovered that one day when I helped myself to two or three of his rolls.

I was staying at my dad's house for the weekend and by the time I got back there, Granddaddy had realized the quarters were missing and got right on my ass. He phoned up, yelling, "Goddammit, I want my quarters!" and "I'm gonna tear you up " and so on.

I was, like, "Granddad, I didn't take the money."

"You know you're lying, boy. I had fifteen rolls of quarters right there on the mantelpiece next to your grandmother's keys!"

For a while I was scared to death to go back over there, but I didn't stay scared long. My granddad stopped keeping his quarters in plain sight. That's when I really started taking.

As I got older, I figured Granddad was kind of like Robin Hood. You know, steal from the rich and give to the poor. I don't think he realized that what he was doing was wrong. When he would take sugar and sweets from the Jewish bakeries, he'd split it up among his four children and make sure that everyone got some. With poor black people back in those days, taking from the white man was acceptable if you used the money to do good for black people. Of course, I didn't have any white people to steal from back in those days, so I stole from whoever let their guard down.

Pretty soon my whole family started calling me "Bad Ass Jeffrey." I went into my aunts' purses at holidays or whenever everybody got together. After a while, they had to lock all the purses in a closet at the family gatherings if I was around because I was such a little kleptomaniac.

Even at that young age all I could think was, "I gotta get mine."

If I was hungry, I'd take money to buy food. And, like I've said, I was hungry all the time. We always had food at Moms's apartment, whether it was government surplus canned fruit, cheese, or peanut butter and jelly. (I hated the government jelly because it came in a can and you couldn't close it once you opened it.) And there were always leftovers that my grandmother would send home with us, so it's not like Moms didn't feed my sister and me. But we craved junk food. We wanted Church's chicken, Jack in the Box, McDonald's, all the stuff we didn't have money for.

My family's financial situation had an even more serious consequence for me. In school, I had problems seeing what the teacher wrote on the board. I was always squinting to the point where my fifth-grade teacher, Mrs. Scarborough, sent me to the nurse for an eye exam. It was discovered that I had problems with my right eye. I could barely make out the letters on the chart.

When my mother took me to a specialist, I was diagnosed as nearsighted and color-blind with a lazy eye. They said I would need eye therapy or else the muscles in my right eye would continue to weaken. They told me to protect my left eye at all times because if anything were to happen to it, I would be legally blind.

We were on Medi-Cal at the time and they wouldn't cover the treatment. It wasn't until I was in my teens and my mother got better medical insurance that I could receive the attention I needed. But by then, there was permanent damage. I wore glasses, and one lens was as thick as the bottom of a Coke bottle.

I looked like a square four-eyed little kid, but I was becoming a very good thief.

I was eight or nine when I started taking money from Moms's piggy bank, a clay monkey head that Cali Slim had made in school. I chipped away at the coin slot with a butter knife until the quarters would slide out. I was in the bathroom trying to shake the quarters out when Moms caught me and gave me a major ass-whopping. She told me, "One thing I can't stand is a liar and a thief." But I continued to get better at being both.

When I was ten, I started taking money from Moms's wallet every Friday when she got paid; she'd usually have five or six twenties in there. Her friend Tom would come over and they'd be drinking into the night, so she never noticed the money was missing. I waited until she took her bath, removed her wallet from her jeans, and slipped out one twenty. Then I'd tell Cali Slim, "I got some money, let's go eat at Jack in the Box."

As I got older, I started getting hungry for things that required more money than my mother's twenties and my aunties' loose change could buy. I got tired of seeing the white boys on TV and up in Northtown with brand-new bikes and nice clothes.

I started getting away with doing wrong. It became a way of life. In South Central L.A., it was a common one—you could be a thief one day and a victim the next.

Back when I was six, my dad showed up for Christmas with a bright green 747 three-speed bike, with gangster white walls, a sissy pole, and chrome fenders. For my sister, he brought a purple one. A few years after Dad split from Moms, he started a whole new family, but he still maintained a relationship with Cali Slim and me. His child support was only $60 a month, but it always came on time. And he always tried to make sure that, when he was able, we had decent clothes and were at his family get-togethers for the holidays.

The bikes he bought us were bad. You know, my dad put those mirrors on them, and they had the little tassels hanging down from the handlebars and reflectors in the spokes. It was like the beginning of the world for me.

That same day, me and my sister were riding our new bicycles down my grandmother's street—five houses down from the Seventy-seventh Street police precinct—and these older boys tried to jack me on the corner. They threw dirt in my face, tried to take my bike, but I wasn't giving up that bike for anything. I just kept yelling and yelling, and pedaling and pedaling. All that yelling must have scared those fools off because I managed to get away safe. Most kids in that situation would have been screaming for their mommy or their daddy to come save them. Me, I was screaming for my granddad—everyone on the block feared him.

I always wanted to be around my father, but I didn't know him like I wanted to. Christmas presents and his rare weekend visits weren't enough. I wanted to play baseball and football with him. I wanted him to pick me up from school, to hug me up and boast about me to his friends. It seemed to me he wasn't any kind of a family man. For him, I thought it was all about the women and us kids just happened to come along.

So my mother was everything—mom, dad, everything.

I remember hearing stories in the family about Moms being molested by a female gym teacher when she was in junior high school back in New Orleans. No one ever told me all the details, but I do know that my mother's family convinced her to go to California and sort of pressed her to marry my father right after she turned eighteen. Blacks in the South were proud people. When scandals came to the family, they always covered them up.

My parents had always had a kind of special relationship ever since they were kids in New Orleans. Mom always stuck up for Dad, who was kind of shy, a humble, reserved man with a skinny build, dark brown skin, and wavy pitch-black hair. My mom lived uptown in a middle-class black neighborhood while my father and his family lived downtown in the crowded Magnolia projects. They met at the corner grocery store where my father worked, making deliveries to wealthy white people in the posh areas of the city.

People who knew my father called him "Little Charlie" because his father was "Mr. Charlie." He was the spitting image of my grandfather, Charles Sr., who gave him the traditional southern upbringing. The good side was: don't lie; don't steal; work hard; be strong. But Granddaddy did not show my dad any love.

For my father, this literally destroyed his soul, his spirit, and his self-esteem. My grandmother kept him close to her during his journey to manhood, always protecting him because he was her only boy. But Granddaddy really fucked my dad up by not showing him love. I don't blame Granddaddy, though, because his father raised him the same way.

My dad was withdrawn, a loner. He didn't play sports or even associate with the other boys in the neighborhood much. His passion seemed to be bicycles and drawing. That's how he and my mother got together—as friends. My dad and his parents had moved to Los Angeles a year or so after the scandal, and he and my mother started talking on the phone long distance almost every night. My mother's family saw it as an opportunity to marry her up, get her out of town and away from the scandal. They persuaded her to go to California and be with my father. Before that, the two of them had never done more than hold hands and fool around.

Once they got married and had my sister and me, the whole family thing became a struggle.

When I turned eleven, my mother's brother Louis was about to go on an eighteen-month deployment with the navy. He told my mother we could live in his house in San Diego rent-free for helping out his wife while he was away. By that time, I was rebelling any way I could think of.

In San Diego, I met a cool white kid named L.D. and we quickly became each other's right-hand guys, exploring the world of breaking into houses, stealing bikes, and other typical adolescent criminal behavior. By thirteen, in 1977, I was an accomplished minor thief. I was failing in school, getting suspended, and not listening to anyone. That's when my mother told my father that I was getting out of control and needed a man's influence. Cali Slim and I went to live with him back in Los Angeles. He was married to his third wife by then, and she had two sons living there, David and Daryl. I think guilt played a major role in my father's willingness to take us.

I attended Palms Junior High School. This white kid named

Fat Tommy—his family owned a liquor store—wanted so badly to be cool with the brothers, to be in with us because we pretty much ran the whole school. He started bringing us money every day that he stole from his father's cash register. Like clockwork, he brought us a twenty-dollar bill and we'd take the money and buy ourselves some extras from the student cafeteria. Then he stopped bringing the money. Apparently, he got busted by his dad.

But that was Fat Tommy's problem. We started putting pressure on him. Like, "What's up? Where the money at?"

The white boy got scared and went to the principal, told him I was forcing him to give us the money. The next time I took a twenty off him, the school counselor pulled me out of class and took me to the principal's office. "Henderson, empty your pockets," I was told. So, I did. "Henderson, where'd you get the twenty from?"

"I been had it. It's mines."

"Well, let's see the serial number."

That's when I knew I'd been set up.

They kicked my ass out of the whole L.A. Unified School District. I was fourteen.

The timing happened to be good, because my dad had just bought his first house in North Long Beach, right on the border of Compton. That location happened to be just the right place to further my criminal career as well. It was where I put together my first crew and earned my first arrest.

It was the summer of 1980, I was just about to turn sixteen, and I was rolling with homeboys we called A-Line, and Marv-dog. We even had a white boy by the name of Ern-dog, who ran with us—he had a curly Afro and he thought he was black. We bought ourselves Dickie jackets with our names on them and we called our little set O-Hood, after the intersection of the four neighborhoods on Long Beach Boulevard. We broke into almost every house in the neighborhood, stealing anything we could carry down the alley—TVs, cameras, guns . . .

It didn't take long for A-Line to get busted by his father. It's bound to happen when you break into the house behind the one you live in. Someone told A-Line's father they'd seen him climbing out the window with someone else. He got an ass-whopping and put on punishment, but I got away with it at first because no one saw who I was. Then A-Line's dad told my dad.

The funny thing was, there wasn't even anything to steal in that house. It had been abandoned after the husband came home and found his wife fucking another man. He killed the guy, pistol-whipped his wife, and went to jail. So there wasn't anything worth looting in the place, but breaking in had a value worth more than money to a teenager.

I was fifteen and I'd never gotten laid yet. There was a girl named DeeDee who lived across the alley from me. We used to fool around, just grinding on each other in the alley and stuff. She started teasing me because I never tried to have sex with her.

She was always saying, "Come on, Jeff. What's up?" She bad-mouthed me to my boys. "What's up with your friend?" she'd say. "I'm trying to give him some pussy but all he wants to do is suck on my tits and grind up on me."

I was unsure, you know, not really knowing how it was done. But then one day I just said, "I gotta do this." I called her up and I took her to that abandoned house and we got busy. Once I knew exactly how to do it, it was on after that—and I had a powerful new motivation to keep ditching school. DeeDee and me started cutting class and going to that house almost every single day.

Even if I was spending most of my time getting laid, there always seemed to be enough hours in the day to fit in a little stealing. Me and my crew were strolling on Atlantic Boulevard one day when I spotted a brand-new ten-speed perched up against a palm tree in Houghton Park. The way I saw it, someone leaves a bike out like that in a park they're asking you to take it. Turns out I was right.

I told my homies, "Look, there's a bike, someone's slipping. I'm getting ready to get it." So here I jump on the bike and I'm pumping and pedaling down the street. Next thing you know, here comes a police cruiser with its lights on, pulling me over.

"Whose bike is this?"

"It's mines."

"Where'd you get it?"

"I been had it. My father bought it for me."

So he takes out the registration number and says, "You know you're lying. This is our bike and we put it here."

Set up again.

At sixteen, I was taken downtown for my first arrest and put in a cell. I was scared, but since it was the first infraction on my record, they let my father come down and get me. They slapped my wrist with six months' probation.

On the way home in my dad's car, he was silent for a while. Then he said, "Did you learn a lesson this time?"

"Yes, Dad. I know I should not have taken that bike. My crew was pressuring me to do it." Little did he know I was lying through my teeth. But Dad didn't appear to be too pissed off. I said, "Dad, what's the punishment gonna be?"

"I haven't decided yet." He ended up grounding me and suspending my TV privileges for a week. The way I saw it, behaving myself for a week was nothing. When it was over, all I knew was that I had to be a much smarter thief. I thought about how stupid I'd been to take a brand-new bike just sitting there in the open and told myself I'd never fall for another sting operation.

The spring semester of 1981 started up; North Long Beach was happening—parties, drama, cars, and gangbanging. And my crew was earning a rep. We were wannabe gangbangers. We only fought when we were forced to protect ourselves.

Crips from Compton used to come to our neighborhood and we would fight them or run—Santana Block Crips, Atlantic Drive Crips, Jordan versus Compton High.

One day I ditched school and got on the bus with my little stepbrother Daryl to go up to the Cerritos Mall about an hour out in white boy territory. I was picking up a shirt for a party that night when we ran into some gangsters from Inglewood.

They were staring me down, mad-dogging me.

That's when they started shouting, "What up, Blood? What up, Blood?"

Bloods identified themselves by wearing red clothes, and the Crips always wore blue. I wasn't part of either gang, but I still had to be careful not to wear the wrong color in the wrong neighborhood. That day I'd made the mistake of venturing out of Crip territory wearing a blue Adidas sweat jacket. All I could do then was just keep staring at those bangers.

I whispered to Daryl, "We're gonna have some problems with these guys."

One guy came up on me going, "What'sup?" I said, "What'sup?"

He tried to grab my bag, but I wasn't letting that happen. In it was a forty-dollar Fila shirt, and forty dollars back then was a lot of money. So we started getting down in the mall right in front of Regal's shoe store and I'm telling my little brother to run. Then I heard a click.

I thought, *I know these fools don't have no gun.*

Somebody reached over me and hit me in the chest. I'm, like, "Fuck!"

Next thing you know these guys back off me and I'm standing there starting to feel this bubbly sensation in my chest.

That click had been the sound of a switchblade.

"Damn," I said, turning to look at Daryl, who was a few steps behind me and looking scared as shit. "Those fools just stabbed me!"

My jacket was suddenly soaked through with blood and my knees started buckling at the sight of it. I began to get dizzy and went down. A crowd had gathered around the fight, but now everyone was backing away as I lay there in a pool of blood.

Some good Samaritan white guys chased down the Bloods in the parking lot while a nurse who happened to be in the mall applied pressure to the wound in my chest with her sweater until an ambulance showed up. I spent three days in intensive care and the doctors thought they were going to lose me for most of that time. My lung had collapsed, there were all these tubes going in and out of me, and I lost a lot of blood.

When I got out of intensive care, my mother was there.

"When you get out of this hospital," she said, "I'm taking you back to San Diego."

I'd been living with my father for three and a half years and I still felt a great distance between us. We just didn't have the father-son relationship I'd been hoping for.

My sister was moving out of his house anyway. Dad's third wife really wanted a baby from my father but he'd decided he had enough children already. My sister, meanwhile, was dating this dude, Money Mike, who had a burgundy Chevy Malibu. An OG (Original Gangsta) hustler from the East Side of Long Beach, he had major game. He'd always give me money and be, like, "Jeff, why don't you go to the store and buy some chips and sodas?" That's when he would have sex with my sister. When I found out he'd gotten her pregnant, I went crazy—but not as crazy as my stepmother.

She told my father straight out, "If I can't have no baby, there ain't gonna be no babies in this house."

My father told my sister she'd have to leave if she had the baby.

So me and my sister packed up our things and moved on to San Diego, where there were better schools and safer turf—or so my mother thought.

THREE

~~~~~~

T-ROW

Moms came to pick Cali Slim and me up at our father's house on a Saturday morning. Marv-dog, A-Line, and Ern-dog all showed up to say good-bye. I was sad to be leaving my friends, but I got excited as soon as we pulled up to my mother's apartment complex, The Condos.

There were niggas on the corner shooting dice, motherfuckers up the block selling dope. I remember seeing a couple brothers hanging out in front of a '77 candy tangerine Monte Carlo lowrider. On the car trunk, a mural of Hobart Street in L.A. depicted a streetlight, a lowrider car, and a naked woman lying on top of it. The two men leaning on that beautiful car and selling dope would soon become my criminal mentors.

They were T-Row and Richie Rich, two Crips who left L.A. because it got so hot for them there, stealing cars and gangbanging. They had the weed business on lockdown in The Condos. We were driving in my mother's '72 Pontiac LeMans with the trunk loaded up and the backseat full of all our clothes from my dad's place. There was something about T-Row and Richie Rich that caught my attention. They had the car, had women hanging all around them.

I was, like, "Damn, it's on up in here!" I couldn't wait to get out of the car and see what was going on inside these condos.

After a couple days or so I was getting up and around in the complex. I went over and introduced myself to T-Row, saying, "Hey man, what's up? This is a helluva car—I like this."

He said, "What's up, youngster? Where you from?" I told him I was from L.A. and he says, "Yeah? We from L.A., too."

T was light-skinned, about five feet, eleven inches, real skinny with a well-groomed goatee and a wavy perm. He was sporting a blue Fila jogging suit and two or three gold Turkish rope chains and a Cadillac medallion hanging from his neck. His pockets were bulging. I was in eleventh grade and he was five years older than me. He was smooth. He had the gift of gab. He was a mack daddy. All the traits I wanted.

T-Row had an apartment that was catty-corner to my mother's, and I would always be looking out the window, seeing all the activity. Twenty-four seven it never stopped, women coming over, thugs shooting craps, partying. And I was always coming over saying, "What's up, T? Can I go for a ride with you?"

He started giving me rides, picking me up from school, and eventually he started warming up to me because he wanted to work me. He could see that I was fascinated by his lifestyle and had the potential to be a great hustler. T never saw me playing the fool with all these other young teenagers. I was serious and attuned to what was going on around me.

We even looked alike. He started telling people I was his little brother, and I took that as an honor. My mother, though, started to worry about my affiliation with this grown man. One day she confronted him in the parking lot of The Condos right under his apartment window.

Everyone respected my mother. They called her Mrs. Henderson. Moms had a rough exterior. She always wore jeans and steel-toed boots because she'd been working as a welder at the Nasco shipyard for the last five years. T-Row and Richie Rich would square up when my mom would come through.

She told them, "Stay away from my son," because she knew the influence they were having on me.

T was, like, "Oh, Mrs. Henderson, we would never put Jeffrey in harm's way. He's a good youngster. We just be kickin' it with him."

Moms told him, "You're too old to be hanging around with him. You know, Jeffrey got in a lot of trouble up in L.A. and this is his second chance. I don't want nothing to happen to my baby. You all need to let him be."

But I wasn't having that. T was the man. I wanted to be around him. I wanted to learn.

He started letting me come up to his crib. T made steaks and fried shrimp all the time. I'd chill with him and his crew, and T would hook me up with a steak from time to time. Then he started having me do little tasks. I'd fuck them up a lot of the time, and he'd say, "You hardheaded, little nigga, you hardheaded. That's what the fuck I'm-a start calling yo ass: Little Hard Head."

He quickly became like a father to me. T-Row might have been a bad influence on me, but he was relating to me like I'd always wanted my real dad to do—just show me some love. I worshipped the ground he walked on. He always had the latest shit, the newest gear, the flashiest jewelry, the finest women. T had a stroll that was so cool you could spot him in a pack a mile away. I began to mimic his walk. Weekend mornings, I'd be out there polishing his lowrider. I'd wax the candy paint job, put Armor All on the tires, polish the chrome rims—everything. I would be dressed and ready to roll, waiting for T to come outside.

When T came out, he'd say, "Yo, Hard Head, what up?"

I'd say, "What up, T-Row? Can I roll with you today? I cleaned the car up and everything. There's not one fingerprint on it and I even swept the dirt up from around the car."

T started taking a liking to me. I felt he cared about me, that he had love for me. Deep inside I knew that he was also playing me a bit. T began to work on my weakness for our friendship and for his way of living. He started using that to draw me in. And that was cool with me because all I wanted was to be his right-hand man. The only problem I ever had with T back at that time was that he'd sometimes give me orders in front of his boys, or, even worse, ladies I was trying to impress. But he saw it as part of my grooming, so I sucked it up as any loyal soldier would.

T began working me hard. He had me doing bitch work, like his laundry, cleaning up his crib. Whatever I had to do to be around him, I didn't give a fuck. It was action, excitement. It was just what I wanted to do. T showed me love, the kind of love I wanted from a big brother or, like I said, a father. We called it "homie love."

T-Row had the weed sack in The Condos, which means he had his workers posted up at all the entrances and exits to the complex.

He used to get "stinky bud"—Thai-stick marijuana—and sell it for ten dollars a gram. He brought it up five pounds at a time from a connection in Mexico. Soon, T had me working as a lookout for his guys who were in the condos selling the weed. While I watched, I learned the game—seeing them weigh it on the triple beam scale, counting the money. T had a spray bottle full of water to add weight to the pot before he bagged it.

Soon enough, T had me out there selling the weed—but only when my mother was at work because he knew if Moms came home and I was standing there with those fools, she would be on his ass. For every ten bags I'd sell, T would give me two. I stashed my cash under my mom's refrigerator, just thinking about having my own weed sack one day.

I was part of a real crew. There was Richie Rich, a former junior college basketball player who was T's obedient right hand; L.A. Will; a dude called Beer Can and his brother-in-law; a couple of middle-class dudes who'd gotten caught up in the game; Roland, a Blood out of Compton who was dangerous because no one could ever read what he was thinking; Omar, who quit his job as a mailman to sling dope for T; and Dino, another Compton hustler.

T looked for guys who were smart, not just someone who could shoot a pistol or throw from the shoulders. Because T saw dealing as a thinking man's game. Even though it was dangerous, you still had to be a critical thinker to be truly successful. You had to have the qualities of a salesman, you had to be a manipulator and a chameleon to deal with a diverse clientele. You had to be able to convince people to buy your product.

Once I started working for T, I usually carried a roll of $400 to $500 in my pocket. That was a big feeling. I'd run down to the mall and get myself a pair of new $100 Air Jordans from the Foot Locker whenever I felt like it. Even when my money was short, like down to $200, I'd break a hundred for a fifty and a lot of fives and ones to make my bankroll look fat. So when I'd roll up on a chick and say, "You wanna go eat?" she'd see this thick wad of money when I paid the bill at Sizzler's Steak House.

I was buying Gucci watches laced with baguette diamonds, gold rope chains, Starter jackets, and Guess jeans. When I'd stacked up $1,000, I took $375 and bought my first car: a '68 Cadillac Coupe De Ville from a white man from East Diego. It was Aztec gold with a white vinyl top, power windows, and pop trunk. I remember the pride I felt just putting in new Christmas tree air fresheners and Cadillac floor mats. Immediately, I went out and stole a brand-new Alpine sound system from a BMW and installed it in my glove box.

With my ride pimped out, I rolled straight up to Skyline and

Meadowbrook, where all the homies cruised and hung out. Every-body was saying, "Look at Hard Head! Hard Head got his 'Lac!" It had huge chrome bumpers, electric black leather seats the size of La-Z-Boy chairs, and big headlights double-stacked in the front.

My mom and sister made fun of the car, though, because it was this big old battleship and here I was this skinny little guy behind the wheel. I think Moms was in denial about where my money was coming from. She let herself believe that I'd saved up enough to buy the car cleaning up The Condos part-time.

T liked my new ride. He told me, "Aw, man, this is cool right here." Then he advised me how to dress it.

I said, "T-man, can I get the rims from the next car we steal?" He eventually hooked me up with some big hubs spoke rims.

I was rolling. I was the man, and you couldn't tell me nothing, even though I was just in the twelfth grade.

T was also having me hold pounds of weed at my mom's apart-ment because it was getting so hot for him. His name was ringing. As long as I was keeping his sacks up in my house, I started pinch-ing off his stash. A few buds here, a few buds there, stacking my own little money on the side.

My feeling was, he was really getting over. Here I was taking penitentiary risks for him and jeopardizing my family's safety as well, and I thought he could be kicking down a little more change for the extra risk me and my family was taking. This was the first distrust I started feeling toward T.

T had a PhD in game. He was a jack of all hustles: a car thief, a master manipulator, and a jacker. He was so fucking charismatic; the women loved him. Even his sidearm showed his style. T car-ried with him at all times a pearl-handled .25 semiautomatic pistol in his back pocket.

He used to have what you call a snatch bar, a tool mechanics use to pull dents out of car bodies, but T used his snatch bar to yank the ignitions out of steering columns so you could start them up with a flathead screwdriver. T and his crew would go out, break into a car, snatch the ignition, and sell off the car in pieces.

One night it was me, T, this guy named Jake, and another one called Black Jesus, rolling in BJ's van out to the suburb of San Diego. T did a reconnaissance mission earlier that week to locate lowrider vehicles that we could scoop by and steal. First on our list was an '81 midnight blue Buick Regal. It was loaded front and back with high-end hydraulics and belonged to this Mexican in one of the car clubs, a Mexican car club that competed against the several black car clubs in San Diego. We got to the location, did a slow drive-by of the house and pulled around the corner. We turned off the lights and the ignition but always left the keys in so we wouldn't be scrambling for them if we had to make a quick getaway.

We crept up to the house dressed in all black, me on lookout while Jake popped the door lock, slid across the seat, and opened the door for Black Jesus. T was posted up in the bushes with his pistol in case anyone ran up on us. After Jake slimjimmed the car door open and popped the cap on the steering column, BJ slid in the snatch bar and pulled out the ignition.

Then T heard a door open and the Mexican runs from the side of the house yelling, "Hey! Hey!"

"Hard Head," T whispers, "get on the other side of the car. Black Jesus, keep going, start the car. We're taking this bitch."

With that, T goes into a crouch position, pulls his .25, and pops three rounds near the Mexican, deliberately missing him—we weren't trying to catch a murder case. The Mexican runs back in his house. Black Jesus starts the car, hits the switches, and the car comes off the ground hissing. T and I run to the getaway car down the street as Jake and BJ peel out in the stolen car.

When Jake pulls up next to the getaway car, T tells me, "Hard Head, get in the Regal—you're driving."

So Jake and BJ hop into the getaway with T while I drive the hot model back to a strip garage BJ had down in the hood. They always put me behind the wheel because I was underage and if I got caught all my mother would have to do was come down to juvenile hall and sign me out.

Soon we were snatching five or six cars a month. They'd get the rims, the music, the hydraulics—and T would give me the fucking rearview mirror.

Eventually, I'd had enough.

"Fuck that, T," I told him. "I want half the money for the rims. Every time we knock a car you guys take the rims and sell the sounds and the rims for like $2,000 and you give me the bullshit parts. We should split everything equally."

T said, "Aw, little nigga, quit whinin'. You should just be happy you get to roll with us. You learnin' the game. Don't you worry, we gonna take care of you. You gonna get yours."

So one day when T was out of town I broke into his apartment and stole a set of hydraulics he was stashing there. Richie Rich was staying with him and someone told him they'd seen me going into the apartment.

Richie jammed me up about the hydraulics. I was sweeping up the parking lot when Richie approached me. He was six feet, four inches, skinny as fuck. He really wasn't intimidating at all, but he was a lot older than me and I knew he had a lot more fighting experience than I did.

He said, "Hard Head, somebody broke into me and T's motherfucking apartment."

I said, "Is that right?"

"Yeah, that's right. And someone told me they saw you coming out with the hydraulics."

"That's bullshit, man. I didn't break into your fucking place."

Richie got in my face, said, "Little nigga, you lying," and pushed me down to the ground.

I got up and started running to my mom's apartment. Richie chased after me but Moms heard the commotion and came out and confronted him.

"What the hell did you do to my son? Keep your hands off him."

Richie backed off, but he told my mother what I'd done and I ended up having to admit it and give back the hydraulics. My relationship with T and Rich became strained after that.

But T didn't take it too personally, because, if he did, he would have fucked me up bad. Deep inside, T knew he was getting over on me. He also knew that I was wising up and that he had to respect the fact that I was growing, that my game was getting sharp.

With a hustler of T's caliber, his name starts getting hot and he has to take a hiatus; so T went on a little sabbatical back to L.A. With T gone, I saw an opportunity to break out on my own. Striking out for myself, I met a connection named Bernard, this older brother who had a girlfriend living in The Condos. Through him I started making my own money, building a client base, and nursing my expanding weed enterprise.

While building my own business, I met this girl Tammy, who visited her sister in The Condos on the weekends. I knew she was digging me because she always sat and watched me through her sister's screen door. Before long, she was sneaking me into her sister's crib on the weekends.

She was a year older than me, already out of high school, and she spent her time working at Burger King while pursuing a modeling career. Tammy was the first girl I ever had a serious relationship with. Even her family liked me. They lived in a nice house in Emerald Hills and her mother used to cook for me all the time.

It wasn't a month before she was knocked up.

At seventeen, with a kid on the way to support, I really had to get my shit rolling. I didn't know anything about being a father because I'd only occasionally been around my real dad. I didn't have any man training and I really wasn't ready for kids. Tammy's mother wanted me to take on some responsibility, so she got me a part-time job as a printing assistant at the printing company where she was a secretary.

At the same time, my weed business was doing good, but word on the streets was that a new hustle was coming down from L.A. When T made his comeback a year or so later, he brought that hustle with him: rock cocaine, as we called it back then, while on the East Coast, they called it crack.

After I "graduated" from high school—they let me walk with a 1.0 and a sixth grade reading level just to get me out of the system—my weed business was quickly going out of style. The wonder drug crack was ten times more powerful than cocaine and ten times less expensive, and becoming the only game in town. It was perfect for poor black folks who couldn't afford the snorting variety, and T and L.A. Will were in on the ground floor.

T started opening up crack houses all over San Diego—in The Condos, at Fifty-fourth and Imperial, on Groveland Street, and on Naranjas Street.

One Friday night T tracked me down at a Jack in the Box that was a local hangout at Euclid and Federal Boulevard.

"What's up, Hard Head?" he said. "How you been doing?"

"I been cool, man. Just getting my grind on."

"I wanna talk to you about some new work I've got going on now. You know I ain't fucking with that weed no more."

I said, "Is that right? What are you doing?" T bent over and went down in his sock, pulled out a little plastic bag with some white stones. I asked, "What's that, homie?"

He said, "Remember when we used to sit back and talk about making millions? This is the meal ticket, Hard Head—rock cocaine. There's gonna be enough for all of us. You wanna get down with me?"

"Hell yeah," I said. "Show me how it works."

That was all it took. I started selling crack out of T's spots.

By then T was dealing almost half of the crack in all of San Diego, and I was just biting off a little bit for myself. Over at Fifty-fourth and Imperial, you thought you were at an indoor swap meet because there were so many people running around there doing business. I used to be there until three in the morning, sometimes until sunrise, with a Safeway bag and one of the homeboys behind me with a nine millimeter watching my back. But, just like old times, I started feeling like T wasn't really breaking me off that much. I just couldn't shake the idea that he was getting over on me again. So, once again, I started telling myself, "Fuck this. I gotta come up and get my own thing going."

L.A. Will and me started a little joint venture on the down-low. I was surprised, not knowing L.A. Will that well, that he would go behind T's back and set me up with an L.A. connection. I figured there'd be something in it for him down the line. After all, it was a cutthroat business, but I didn't give a fuck. I just wanted the L.A. hookup.

My business was so good that working my legit job at the printing plant got difficult. I gave my two-week notice to the printing company and devoted all my time to the crack game, secretly selling my own crack at the same time as I dealt T's.

By 1984, T was on his way to becoming the Godfather of San Diego, but he lost his temper over some punk-ass shit. T caught a case and had to do some time. He'd gone to a car show in L.A. and a guy he'd had a beef with back in the day stol on him—that is, he sucker punched T and knocked him out. T gets up, goes and gets his .25, comes back and blasts the guy. That got him a few years in the pen. Luckily, he only grazed the motherfucker.

With T gone, I had no choice but to establish myself as an independent player and move out of T's shadow.

Meanwhile, a lot of drama started heating up around Fifty-fourth and Imperial. This guy Frank "Little Man" Brown—an East Side Crip from L.A.—opened up a crack house there. Rob-dog had an operation there, too, and so did Schoolyard, a Blood out of the Syndo mob. With all the rivalries, the drive-by shootings soon brought the whole business to the forefront for the Feds.

The danger really didn't bother me, though, because I was neutral and didn't have a beef with any of those fools. I was still buying crack from L.A. Will, but I had put together a crew of my own. As the months went by, my dollars were stacking and I was able to put a few homies in business, by fronting them packages on consignment.

I was becoming the man.

FOUR

~~~~~~

MY FIRST ROLEX

I became so successful that L.A. Will couldn't keep up with my demand. He called his L.A. connection and asked if it was cool to introduce me directly. Will saw it as an opportunity to get his own supply more cheaply by bringing his people a new customer with great earning potential.

So Will tells me, "I'm-a take you to L.A. and introduce you to these guys in me and T's neighborhood called the Twins."

"What do you mean the Twins?"

"They're twin brothers. They're ballin' out of control on a major level."

One Saturday morning Will scooped me up in his blue rag '66 Chevy and we bounced to L.A. The Twins lived around the corner from where T had grown up. I was a little nervous at first, but L.A. Will assured me that everything was going to be cool. We rolled up in front of this green stucco house with a well-manicured lawn, and one of the Twins came out the front door while his brother came around from out back.

I thought maybe they would be some gangsters or something, but these were some humble and square-looking brothers sporting

L.A. Dodgers caps. I was shocked that they were selling dope out of their mother's crib. Will had told me that they were plugged in with some of the top cocaine dealers in L.A. They were the buffer between the brothers and the Colombians.

In the driveway, I noticed they had some nice cars—two convertible 300 ZXs and a convertible 450 Benz Coupe.

We had the introduction thing going on, sizing each other up, then we walked into the backyard where there was a room attached to the garage in which they kept their stash, which doubled as a seamstress shop where their mother would design and tailor clothing for her handpicked clientele.

One of the Twins asked me, "What are you trying to do?"

I said, "Man, I need to get some work. I used to get it from T, but—"

"Oh, T's my homeboy, we grew up together right here! We know T."

"Well, I've been up under T for some years but, you know, he got knocked."

"Yeah, yeah," Twin said. "So how much do you get on a regular basis?" He was trying to figure out what kind of business I was doing down in Diego and what the demand was like, just like any intelligent businessman.

"I want to start out with an eight ball, and then we'll see how it goes from there. I'll take that back, see how the smokers like it. If it's good, I'll be back for more."

The Twins said that was cool.

One of the Twins went into his pocket and showed me the work. It was a medium-sized, cookie-shaped rock with several lumps. The Twins used to rock up their powder one ounce at a time in mayonnaise jars. After the cocaine gel hardened it would look just like the praline candies with pecans that they sell in New Orleans.

Peeling off $300 from the bankroll in my pocket, I said, "I like this, looks good," and we exchanged pager numbers. Me and Will jumped in the '66 Chevy and headed back to Diego.

I was like, "Will, man, how'd these guys get hooked up like that?"

"It's all about relationships," he said. "They established solid connections with these guys who have the South American connection and they put 'em down with the birds."

"Do they buy it?"

"Nah, man. They get it on consignment, like a hundred birds at a time."

I said, "That's what I gotta do. I gotta increase sales and come up with a slick marketing plan to get all of these Diego niggas buying from me—but I don't want nothing on consignment. I don't want to work for nobody no more. I'm doing my own thing. Now that T is gone, I'm calling my own shots."

"I know what you're talking about," Will said. "But you know, right now I gotta keep T's sack working. You know, hold down the houses until he gets back."

Will wasn't a thinker. T kept him in debt so that Will was obligated to him and was afraid to try to break out on his own.

The first thing I did when we got back to San Diego was to head straight to the crack house on Naranjas. The house still belonged to T, but Will let me come in and move my package.

I broke off a little chip from that eight ball and gave it to this smoker named Funky Blood. He was a young Blood gangster from Skyline, the neighborhood where I would eventually establish my home base. Funky Blood was a skinny motherfucker who always had a white film around his mouth and wore a beanie on his head because he never combed his hair or bathed. Seemed like he never slept, either. You'd catch him coming out of alleys and hanging on street corners all around Naranjas and Fifty-fourth and Imperial.

But he was the type of guy I always kept around because he kept his ear to the streets—always letting me know who was snitching when the police came snooping around, who came through the crack house, who left the crack house, who did what in the crack house when I wasn't around.

So me and Will go into the spot to test the Twins' dope on Funky Blood. He fires up in his glass straight shooter and hits the pipe until the rock disappears in a cloud of smoke. Funky Blood started licking his lips and pacing the crack house, talking fast, and praising the dope.

He tells me, "Damn, this is some good shit. I can help you move this."

I said, "Oh yeah? You like that?"

"Yeah!"

I knew at that point that I was going to have shit sewn up in Diego with that product. After that, I told Funky Blood that he was going to be my personal crack-tester. I started letting him keep a $50 rock for himself just for testing out my shit. As fucked up as he was, he was as valuable to me as anyone else. I never smoked crack, or did any drugs, besides taking a onetime hit on a joint with T. I barely ever drank. Some of the other homies were closet smokers, but for me it was all about the money. Nothing was going to stop me from stacking my dough. A guy like Funky Blood, though, was necessary because he was quality control. When I didn't have someone like him around, I was forced to taste the dope by touching my finger to my tongue. I hated having to do it; the closest I wanted the drugs to me was to be coming out of one pocket and going back in as green paper with dead presidents' faces on it.

The rest of the eight balls sold out in less than two days and I was ready to re-up with the Twins, but I wanted to wait until the weekend to make the trip. Back at Moms's house, I counted out all my loot. I had about $9,000 I'd saved up over the last six months.

When I paged them that weekend, the Twins called back within ten minutes.

I called them back from a pay phone. "What's up, Twin?" I said—from that point on, I always called them Twin, no matter which one I was talking to. "This is Jeff, you know from Diego. I met you with Will about a week ago."

"I'm listening," Twin said.

"Your shit was good," I told him. "Listen, man, this is what I want to do: I want a bigger package this time around because I don't want to be running back and forth, up and down this free-way. That doesn't make no sense, it's too risky. I want half a bird this time."

"Yeah, okay. We can do that."

"What's the price on that?"

He said, "I'm-a hook you up— 8,500 for half a key, and the more you buy, the cheaper I'm-a give it to you."

"Twin, I hear niggas be dumping them for like $7,500 or eight Gs."

"Jeff, our shit is good. Motherfuckers will be robbing and jack-ing to get this Rolex dope. We stand by our product."

"Bet, Twin, let me see for myself. Can I come up in the morning?"

The next morning, I jumped in my '78 root-beer-brown Caddy and bounced to L.A. I went by myself this time because I was hop-ing to get in some good conversation with the Twins away from Will. I wanted to get to know these guys, figured if I could gain their trust maybe I'd get them to come down on the prices. I felt that the Twins wanted to build a relationship with me because they knew Diego was a hot market and they didn't have anyone

working it. Will was only a part-time buyer and T had other connections.

So the Twins needed me as much as I needed them.

They had me meet them over at one of the Twins' girl's house
over in Hawthorn. It was a busy street, a lot of traffic. Twin came
out to meet me.

"What up, Jeff?" he said.

"Not much, man, just trying to get it going on a major level
down my way."

"Let's roll upstairs."

When we got inside, I figured this was one of the stash houses
where they cooked their dope because there was paraphernalia all
over the place—Ziploc bags, glass pots next to the stove, and boxes
of baking soda on the countertops, but I didn't ask any questions.
There was a fine sistah stretched out on the sofa, and I didn't ask
about her either, but I sure as fuck was watching her out the corner of my eye.

Twin went to the kitchen and brought a triple beam scale out
from under the sink—a nice one, too: chrome plate with a large
dial for adjusting the weight. After he weighed up half a key, he put
it in several Ziploc bags, put those in a brown paper bag, and then
put that into a Foot Locker bag—niggas always had Foot Locker
bags around from the Air Jordan sneakers we bought.

I peeled $8,500 off to the Twin in $100 increments, leaving
me with just $75 in my hand.

"Here goes your change," Twin said. "I'm going to knock off
$500 for you." Then he offered me something to drink and asked
me to sit down at the table. "Who's running things out there in
Diego?"

I said, "Man, there are a few cats that come out of L.A.; one
of them is Frank 'Little Man' Brown from the East Side. We got
some crews outta Long Beach and some other wannabe Gs coming

down there setting up shops. But, man, I'm getting ready to lock things up down there. Everyone loves your shit. It's good."

"Whatever you need, we be down with you. We might even look at storing some weight out there or matching you one-for-one on anything you buy," Twin said, and then paused and looked down at the stack of money on the table. "Let's see how things go."

"I'm going to be back at you two or three times a week re-uppin' until I'm ready to get the big sack. I just need to know that you can provide me with the work all the time and that the quality of the product will stay the same."

"Aw, homie, our product is good. We only buy the straight Peruvian flake. You ever seen a whole key?"

"Only on TV."

He motions toward the sistah on the couch, and she gets up and goes into the other room. When she comes back, she brings to us a brick-shaped package that looked like it was wrapped in a castlike material, like doctors use on broken bones, with a strip of duct tape wrapped around the package marked "Rolex." Twin took a knife and cut it open. Then he showed me these thin, flakey wafers of cocaine.

"This is the level you want to get on," he said. "Buying these."

I was, like, "I ain't never seen anything like that before. I've seen little eight balls, and sixteenths, and that half key of rock, but I've never seen this much powder before. That's a trip."

Twin walked me outside to my car and I put my package under the spare tire in the trunk and drove back to Diego. As I cruised that two-hour trip, at the legal fifty-five miles an hour, I started dreaming . . .

Damn, I have to get on the level of these guys. I like how they do business. They're not no gangsters. If the police saw them walking down the street or driving, they wouldn't bother them.

Even though I was a hustler, I'd never do some of the stuff I'd seen T and some of those other niggas do. I didn't smoke weed, I didn't drink, and I rarely carried a gun. To me, it was all about the money, status, and helping my family out of poverty. I was on the paper chase. I didn't care if they were Crips, Bloods, or East Coast niggas. It didn't make a difference to me. I just wanted to sell my product and stack my bank.

And that's just what I did when I got back to Diego.

Even from prison, T still called the shots at his crack houses on Naranjas, Fifty-fourth and Imperial, and over on Groveland Avenue through his L.A. crew. All of those spots were getting hot. They were having police raids all the time. Will and them were flushing dope down the toilet, jumping out of windows, and paying more money out to crackheads to be lookouts for the police. I knew I couldn't sling out of those houses anymore. It was too risky.

I got smart. I quit selling the rocks to smokers and two-bit dealers and started selling the weight. I let everybody know I had ounces for sale (twenty-eight grams), and I always added an extra gram or two to fatten the package up a little bit so that everyone would buy from me. I stopped being on the street; the only way for a client to reach me was through one of my pagers.

I started supplying all the midlevel dealers in most of the black neighborhoods. Before long, even buying half a key at a time, I ended up having to drive to L.A. three or four times a week. After a month or two, I moved up to buying my first kilos.

Twin said, "Look, Jeff. You been buying a lot of weight from us, you're one of our best customers now. From here on out, if you buy a key or more we'll meet you halfway between L.A. and Diego. We'll bring it or we'll have one of our boys bring it."

Then I worked up the nerve to ask the Twins about a consignment deal where they'd front me two keys for every key I bought because I moved them so fast that I just couldn't keep up with the demand. It was selling too fast. After four or five months I was moving eight to ten kilos a week. I had to start bringing the Twins fifty- and hundred-dollar bills because the money was getting to be too much to count and harder to stash. Imagine trying to buy ten kilos at $17,500 a pop with fives, tens, and twenties.

I wised up about my business. I began to curb back how much I'd hang with my crew in public. I mostly saw them on Sundays, when most dealers shut down shop, because it was the Lord's day. So Sundays were lowriding days. We'd pull out our best cars and cruise Crenshaw Boulevard in L.A. or hang out at Jack in the Box in San Diego—profiling and showing off our self-worth.

I told most of my homeboys I was spending more time with my son, but I was really on the move around the city, hooking up with new clients who were enemies with some of my other clients, doing some marketing, some PR work—you know, letting smokers from different neighborhoods try out my dope so I could get the buzz out there on the street that I had the best shit.

I added to my car collection. I bought an '86 Chevy Mini-Blazer from the Twins for $15,000. I put the candy paint on there and some brand-new chrome rims. With my two Caddies, I had three cars now.

The time had come to move out of my mother's place. I was making too much money and was always coming in from the streets at all hours. Moms didn't like that too much, and I was really trying to keep her out of my business. Besides, I had been working my sister into things—having her do L.A. runs to pick up from the Twins and taking care of my personal errands—so I didn't want Moms to take notice.

Moms looked the other way regarding my criminal activity.

Hell, after struggling most of her life to raise my sister and me after Dad left, she had grown tired of the system and had lost her faith in the American dream. White men were giving her the blues at the shipyard. Moms often confided to me that one day those white boys at the shipyard would pay for her hell. Being one of the only women on the job, and the only black one, they would piss in front of her and make racist jokes and sexist remarks. She often broke down in tears. I even offered to have my boys pay them a visit.

She said, "Don't. I couldn't live with myself if something happened to you."

Even though Moms never openly condoned my lifestyle, she began to change, indirectly going with my program to get our family out of poverty.

My first crib was a one-bedroom in a condo complex down in the posh suburb of Bonita Vista for $650 a month. It had a living room, a dining room, and a garage, where I rented two spaces for $150 a month. I bought a gray crushed-velvet sofa and loveseat set, a glass dining table with blue velvet chairs, and one thing I'd always wanted: a king-size waterbed with a motionless mattress and wooden headboard with mirrors. I couldn't wait to have one of my girls come over and kick it with me on that bed.

After about six months, I came home one night and saw a German shepherd leashed to the stairs that led to the condo above me. It was a K-9. The fucking police had moved in upstairs. I didn't think they were staking me out, but I still had to raise up out of there because I didn't want to take any chances. So I found a place out in Chula Vista, another posh enclave, for $1,000 a month, and bought a whole new set of furniture. This crib was phat. It had a one-car garage under my unit, a washer and dryer, a fireplace, and a floor safe in the closet were I kept a small amount of petty cash—not more than ten Gs—and some jewelry.

It was the summer of 1984; I was twenty years old and the head of my own crew.

My homeboy Hump was an OG Skyline Blood who'd done some time in the California Youth Authority. Hump was a gangster but very reserved; he didn't fuck with too many guys. Hump's brother was Jake. We used to call him "Jake the Snake" because he was so scandalous when high. He was an alcoholic and a closet smoker, and he'd been T's number one car thief. Stevie Bobo was a straight-up smooth player. Stevie and me used to compete all the time about who was going to get with which girl, but he was still living with his mother, trying to get bank and represent like the rest of the crew. Black Blood was a helluva auto mechanic. He was so black that at night all you could see were the whites of his eyes. He always dressed in all red, and every car he owned was red—he truly represented the Blood card. Troy, who we called Silky Slim, was real skinny with little bird legs, but he was swol' as a motherfucker in his arms and chest because he'd trained as a boxer. His brother, who died with thirteen other members of the U.S. Olympic boxing team when their plane crashed near Warsaw in 1980, had trained Silky, so he was the fighter in our crew. The Hitman was about six feet, three inches, real buff, and had been in and out of prison all his life. The Hitman was what he was. He wasn't a regular part of the crew, but we always took him with us when we went to concerts and public venues because the Hitman would look at you and cross them arms and wouldn't no one dare approach us.

Last but not least was Tank Bo Bo—a short, chubby Blood but solid as a rock—who'd been gangbanging all his life. Tank was a closet crack smoker when I met him, always fidgeting and sweating, and his thumb and index finger were always burned. He became my bodyguard and right hand. I saw potential in Tank Bo Bo like T saw in me. His drug use was a problem, but Tank had a serious rep and was a good hustler. So I decided to invest money and time to build him up.

I usually avoided concerts and house parties because that's where all the drama always went down. I dealt with Bloods and Crips from all over the city and I didn't want to get caught up in any rivalry bullshit. But when Hump came up one day and told me he had tickets to a Whitney Houston concert, I thought it might be a good opportunity to get out and do a little PR with that good Rolex dope I had.

The whole crew met up at the Jack in the Box at Euclid and Federal just before the concert, jumped on the Martin Luther King Freeway, and headed to the San Diego Sports Arena. I was leading the caravan in my new fully dressed pearl white 500 SEL Benz with Tank sitting next to me. Hump was behind me in an '82 candy apple red Fleetwood Brougham Caddy; Black Blood was in his '78 Fleetwood, and we were all riding two deep. We had one of the homeboys, Pooh Ru, trailing us in a nondescript car with some heat just in case, Pooh Ru was a crazy youngster, who was a die-hard gangster and never minded putting in work.

We were clean, had fresh done-up Jheri curls and perms; everyone was draped up in their 14k gold rope chains and gold nugget ring with diamond chips and sporting top-of-the-line alligator and suede shoes.

The arena was packed. Security was tight with SDPD officers everywhere, and I could sense the undercover detectives all over. The air was thick with bud. All the different crews were taking up their own sections. Girls were posting up in little cliques, choosing and trying to get chosen. Everyone was showboating.

A lot of guys and females I knew were in there, and many of them didn't particularly care for myself and my crew. So I became very security conscious, constantly observing, taking reads off people. Immediately, I started pressing up on potential clients.

I saw the East Diego crew: CS, his little brother, and some

other cats that ran with them. Their whole neighborhood was all low-income apartments and no houses, so it was easy to set up crack houses there because there were so many alleys in which you could always get away from the police. The East Diego crew was sitting on a cash cow and they didn't even know it.

So I say, "Yo, CS—what up, homie?

"What up, Big Jeff?"

"Not too much, man."

All the crew members head-nodded one another, my crew behind me and CS's boys backing him.

"Check it out," I said. "You know I got some of that good shit, man. Some straight Peruvian flake dope."

"That's what I hear."

"I got a money-back guarantee. It comes already hard, you don't gotta worry about cooking nothing. It comes already chipped up and bagged however you want it."

CS said, "Yeah man, I hear your shit is real proper. I've been wanting to get with you for some time now. I'm-a call you tomorrow."

"Cool, homie. All you gotta do is put it on the triple beam, weigh it out, and you're all good."

"Bet," CS said and we exchanged pager numbers.

Strolling the concert venue, we found a little wall and posted up, watching the girls go by. One of our homies from Skyline named Big Hollywood rolled up on me, saying, "What up, Hard Head? Somebody wanna meet you."

I said, "Is that right? Who's that?"

"This girl named Carmen."

"I don't know her."

"Oh, you'll know who she is when you see her."

"Tell you what," I said. "I'm-a give you my pager number and you hit me up when you track her down."

Even though I had a little family (my son Jamar was three years old), I was still always interested in other women. Tammy knew I had several girls on the side. She didn't trust me at all. A couple months back, she'd found two movie ticket stubs in my pants and cut up all my clothes with scissors. But I think she hung in there because she believed she could make me change my ways.

About halfway through the concert, Whitney takes a break and I run smack into Big Hollywood buying brewskis with this fine-ass girl.

He said, "What's up, Jeff? This is Carmen."

"Hey, Carmen, how you doin'?" I said. Then I pulled her away from the crew to have some small talk with her. "I seen you before. Don't you live down there in Lincoln Park?"

"Yeah."

"Over there on—what's that street you live on?"

"Olivera Street."

"Yeah, that's right. You go wit' that dude Harold."

She said, "Yeah, yeah, you know . . ."

"That's all right," I said. "You don't have to explain."

I looked her up and down, like, *Damn!* She was thick, she had back—smooth brown skin, her short, wavy hair whipped up tight, close to her head. She was very, very sexy. I couldn't wait to get with that. We exchanged phone numbers and I thought, *I have to have this! And she's choosing, too!*

After the concert, Tammy called and said, "You comin' over?"

I dropped Tank at his crib and went to kick it with my little family up on the Brook, the Meadowbrook Apartments up in Skyline. Tammy had a two-bedroom Section 8 apartment where she lived with my son. But before I even got into the apartment, my pager was blowing up. Damn! It was Miss Carmen.

Two o'clock in the morning and she must of wanted one of them late-night booty calls. When a chick calls you after midnight, you know what that means. But I didn't call back because I was already sliding the key in Tammy's door.

On Sunday morning, I decided I wasn't going to do any hustling. I turned off the pager and told Tammy, "Wake Jamar up, we're going to L.A. and do some shopping. Let's spend some money."

We were in Tammy's bedroom. She was getting Jamar dressed when she started hitting me with a lot of questions: "Why haven't you been around lately? What you been doing? I know you're out there making your money, but what about your son and me? You need to spend more time with us. I know you been fucking around out there, Jeffrey."

I just told her, "Baby, I'm doing this for us. I'm trying to make sure our son has everything he needs, that you have everything you need."

I loved my son, and I loved Tammy, but I wasn't letting anything keep me from my dream to move us all away from the ghetto and get my mother out of that shipyard—put money in her pockets, get her out of debt, and send her on plush vacations. But Tammy wasn't down with that.

"What are you talking about?" she said. "I don't give a shit about that stuff. I don't care about the money, Jeffrey. I don't care about those cars, none of that shit. I just want you to be there for your son. I want my son to have a father."

"I'm making money for us," I said. "You've gotta understand that. A lot of the females out there, I'm just doing business with them. They're helping me get paid by taking penitentiary risks for me. But I'm not fucking around with none of them."

She said, "Jeffrey, that's bullshit. I don't believe that. Listen, I don't want to talk about this shit in front of Jamar."

We went to the mall and spent several thousand dollars. I even bought Jamar a tailor-made Michael Jackson leather suit. Around six in the evening, I dropped off my family and rushed to call Carmen. She'd been on my mind all day.

"What's up, Mr. Jeff? I been calling you."

"I know. You know I was real busy today, some major shit I had to do."

"I called you after the concert."

"Is that right? I must've just fell asleep." I was lying through my teeth.

She was, like, "Yeah right."

"I did! Anyway, what you up to today? Can you come over tonight?"

She said she would, so I turned on the fireplace and started cleaning up the crib. I really wanted to impress Carmen because she was a young tender and fine as hell. I'd always had eyes for her when I'd see her hanging on her block but I'd never had an opportunity to get at her until now.

When Carmen showed up, she was looking good and smelling good. She had on a little baby-blue Fila tennis skirt and matching jacket with the zipper down just far enough to show off her cleavage. "Wow," she said, "your place is nice!"

"Yeah, it's okay, but I do need a woman's touch on it."

"Well, I can help you with that."

"Is that right?"

So I take her to the couch, grabbed the remote, and put on BET. After we watched a few videos, I gave her a complete tour of the crib. Then we sat back on the couch and I put a Keith Sweat tape in my stereo—"Make It Last Forever" was the hit. We started making out. I put my hand on Carmen's knee and I was kind of surprised that she didn't offer any resistance when I moved my hand up those smooth brown thighs. Before you know it, we're on the floor in front of the fireplace and the Keith Sweat tape had repeated two or three times. By that time it was on, we were both really feeling each other, and we made love for the first time.

We fell asleep holding each other. I slept very hard that night and woke up to bacon and eggs, cheese grits, and my favorite: Eggo waffles heavy with butter and that thick-ass maple syrup, just like at Grandma's house.

I said, "Carmen, I didn't have that stuff in my fridge."

"I know. I wanted to make you breakfast so I went shopping."

Damn, I thought. *That's what I'm talking about.*

While I fed my face, Carmen became a domestic diva, cleaning my whole place the right way. Just like that, Carmen started caring for me; it seemed like she knew me. The way she was looking after me reminded me of my grandma waking up early and cooking breakfast for me and my sister and taking care of the house. It was something I'd always wanted at my mom's place. Carmen was really special. She was in.

After an hour, my house was clean and I was on full. Carmen's pager started blowing up.

I asked her, "When can I see you again?"

"When do you want to?"

"I wanna see you later today."

"All right. Let me take care of some things at home and I'll call you."

Over the next several months we saw each other almost every day. I was amazed by her passion for me—passion to see that I ate right, had clean clothes. I knew she would be my main girl, but I still had unfinished business with Tammy. I'd been drifting away from Tammy for a while now, and there was no way I could be with her and Carmen at the same time.

The night I went to break things off with Tammy, I noticed

a yellow Citation parked in her space. My heart started racing. I
didn't think I'd be jealous. I called the house on my new car phone.
She didn't answer, so I went to the door and started banging on it
until she finally yelled, "Who is it?"

I said, "It's me. Open this fucking door."

Next thing you know, I see the yellow Citation peeling out of
her parking spot.

"Tammy, open up this door!"

"Jeffrey, it's late. What are you doing here?"

"I want to see my son." She opens up the door and I say, "What
the hell is going on? Who was the guy in your parking space?"

"Nobody was over here!"

"This guy was in your parking lot. You were in there fucking
him!"

She swore up and down. I went past her to my son's room
and saw him sleeping. After I'd kissed him and hugged him up,
I went back out and asked Tammy what the fuck was going
on.

She said, "I don't love you anymore."

I went crazy and knocked over the mantelpiece. Everything
crashed to the floor and my son woke up.

"What do you mean you don't love me? How can you just walk
away from me?"

I was fucked up, devastated, but it was my pride more than
anything—especially because it was some broke-ass motherfucker
driving a Citation that could take my girl from me. I'd thought I
had her heart on lockdown.

So I left. Back at my place, I called Carmen and told her to
come over. She said she didn't have a ride so I told her I'd come
and get her. She made me pick her up around the corner from her
house because she didn't want her family to see me.

"I've been with Harold for a few years and my family knows

him very well," she said. "I still want to see you but I have to do things right. I need a week or two before everyone knows we're together."

Over the next few months my business thrived. Things were really big. I stopped buying dope hard; I needed it soft so I could blow it up myself and enhance my profit margins. Plus, the Twins had gotten too big. They had stretched their crack line across America and I had to wait too long for them to cook up my orders. Big dealers in L.A. were selling the recipe to blow up crack for twenty-five to thirty grand, and you could make your money back with two buys, but I didn't want to spend that kind of money. Besides, I'd been observing the Twins and Will cooking rock for a couple of years and I knew I was ready to get in the kitchen and do it myself.

I bought one bird of powder from them for $17,500, bought my cooking supplies at Kmart and rented a room at the Spring Valley Motel 6. I experimented by cooking in small batches at first, just in case I fucked it up. First, I weighed out eight ounces powder and four ounces baking soda and premixed them in a salad bowl. I brought my bottled water to a boil, just like I saw the Twins do.

I wasn't sure whether the water was supposed to be boiling or simmering before I added the dope and baking soda, so I lowered the water to a simmer and added the mix. I waited nervously for the ingredients to gel. As it began to gel, I felt a little relief. I hurried the glass pot to a sink full of crushed ice. It immediately turned into crack.

I was, like, "This is the shit!"

Once I removed the small crack plate from the pot, I blotted it with a dry towel and placed it on the triple beam scale. My eight ounces of cocaine yielded a return of twelve ounces of crack. Selling crack at $1,500 an ounce, those extra four ounces would give me a profit of $6,000 per half bird. That meant I'd make $12,000 on every key I bought, and I could easily move five and ten kilos on the first and fifteenth of each month (which were the welfare paydays).

I thought, *Damn! I should have been cooking my own shit a long time ago.*

By the time T resurfaced from prison in '86, he was near broke and I was hood rich. He'd left Will with his business, and Will had fucked it up and was now in the hole with T. T confiscated all of Will's vehicles and his jewelry and started treating him like a bitch. I felt bad because Will had helped me get ahead, but that was the game.

When I heard T was out, I went to T's mother's house to see how he was doing. As I pulled up in my new custom white 500 SEL ragtop Mercedes Benz, I saw T in the yard with a couple of his homeboys.

"Welcome home," I said. "Damn, T, you look good." He'd buffed up in prison. He was packing some eighteen-inch guns and a ripped chest.

"Damn, Hard Head, I been hearing about you while I was away, that you been rolling and got San Diego sewed up. So tell me, little brother—can your big brother come down there with a sack and put it down? You know them niggas fucked off my money and my spots."

T had gotten some keys on consignment from some guys in his neighborhood, but he'd been away so long that he had no clientele. He wanted me to move his sack for him.

"T, you know I'd do anything for you. Give me a few days and I'll get your money for you. Matter of fact," I said, taking out my roll, "here's five Gs."

It was customary to give a friend some money when they got home from prison without having them ask.

"Now let's hop in the Benz and go to the mall," I said. "You feel like doing some shopping?"

"Bet."

I took him to an L.A. store that he'd turned me on to years ago, which was owned by some Iranians who took care of all the dope dealers on the fashion end. I bought him a couple of dress suits, some sweat suits, shoes, and a leather jacket.

My relationship with Carmen was in full swing. Everyone in Diego knew we were together, but I still had to play down the relationship in public because my girls on the side were causing me some small grief. Girls who I used to transport and stash my dope were sweating me about not coming over to their places as often, not giving them time on a regular basis.

I rented Carmen an apartment in Spring Valley not too far away from my home. I'd made sure it had a gas stove because I was slowly bringing Carmen into my operation. We went to a furniture store and I spent $2,000 getting her place set up. Carmen was becoming a top-notch bitch in San Diego. All her friends were calling her, wanting to hang out. She was wearing all the latest fashions and sporting the newest imported purses. I was draping her in diamonds and gold and she was getting her hair and nails done at least twice a week.

I felt confident with Carmen at my side. We were like Bonnie and Clyde. We ate together, went to bed together, traveled together, and hustled together. I started giving her crack cooking lessons. She was a fast learner. She was a good home cook as well, so she was a natural in the kitchen. Her mother had taught her to be very domestic and family oriented. I trained Carmen to cook the dope, weigh it up, bag it, make drops, and collect money.

With Carmen as my partner in crime, I could relax a little more. If I was out networking, hustling up new clients, I could call her up and say, "Baby, go ahead and cook up one of those birds—I've got an order." Then I'd roll over to her house and everything would be cooked, bagged, and ready for distribution.

I bought Carmen a brand-new gray 1986 Nissan Sentra, a straight undercover car—no rims, no music, and no tints on the windows. She'd drive me around making my drops. But no matter how tight we got, I could never quite get her family to accept me.

Carmen had two sisters and seven brothers. Her mother was a nurse's aide and her dad hung drywall. Even though two of her brothers were closet smokers and her older sister was dating a dealer, her siblings still thought Carmen could do better than me. She'd had a good guy before I came along.

One night, Carmen and me fell asleep at her parents' place and her father found us in bed the next morning.

"I don't mind you coming over to see my daughter," he told me. "But don't ever disrespect me or my family. I don't want you to spend the night at my house."

"Yes sir," I said. "I apologize. It just got late and I fell asleep."

"Carmen," he said, "I don't ever want this to happen again."

"Okay, Daddy. I'm sorry."

In April of '86, Carmen's brother arranged a surprise party for their mother's birthday. Carmen's parents had never been on a trip together. So I went to the Eastern Travel Agency—the only black-owned travel agency in San Diego—and bought her parents round-trip tickets to Hawaii, a seven-day stay at the Rainbow Hilton, plus $1,000 spending money.

When Carmen's mom was opening the gifts, she said, "Oh my goodness! Tickets to Hawaii!"

But one of her brothers whispered in her ear that she shouldn't accept the gift because the tickets were bought with drug money, and they never took the trip.

Aside from my problems with Carmen's family, my life was unbelievable. I was making so much money I didn't know where to put it. T's protégé, Little Hard Head, had taken all the lessons learned over many years from T's schooling. I had game, respect on the street. I had a fleet of cars, the flagship being my white custom 1985 500 SEC convertible Benz, fully dressed with the Lorenzo rims, Alpine sounds, paint, and car phone.

I was one of the youngest players, but one of the richest, too. There were women, first-class trips to Jamaica with my crew, and to Maui with my women; shopping sprees on Rodeo Drive; ringside seats to every big title fight in Vegas.

I purchased a trilevel custom-built home on Dictionary Hill: sixteen hundred square feet, three bedrooms, two and a half baths, plush carpeting, and a Universal gym. I added Italian custommade furniture, Sony Trinitron TVs in every room, beveled-glass mirrors all over the walls, all the latest appliances from Sears, and custom Kohler sinks and toilets. I had the yard landscaped with palm trees and electronic sensors built into rocks that shot my sprinkler system into action, and a brick wall around the whole crib with a wrought iron fence at the driveway. There was even a dog run for my rottweiler, Gangster. All that and I had a quarter million in my closet.

I was twenty-two and untouchable. I was becoming relaxed and flamboyant. Mostly just sitting back counting cash, I was living the American dream. I knew that there was a major demand for my dope, but it never dawned on me how many crack addicts were filling the hood. It was just money to me. At times I'd get nervous that it would all come to an end, that somehow I'd slip up, but those thoughts would come and go, even though many people warned me to slow down.

My sister, Cali Slim, was dating a San Diego police officer. She worked very hard to make it appear that she had nothing to do with my criminal activity. As far as he knew, she made her living working at the front desk of a Marriott Hotel. He used to get in her ear all the time: "You need to tell your brother to slow down. His name is ringing in the police department."

When my sister told me, it just went in one ear and out the other.

FIVE

~~~~~~

CAUGHT UP

I had eight cars, each one worth more than $30,000, and it still wasn't enough. Things were really getting out of control. When you have that much money, people start to pay too much attention.

One time the Twins sent one of their workers to meet me with ten kilos, but someone who'd been at their crib when I'd placed the order had overheard the conversation and followed the delivery-man all the way down to San Diego. While the deliveryman was waiting for me at a motel parking lot in Bonita Vista, one of the Twins' boys put on a ski mask and jacked him at gunpoint.

Everything was chaos. Weeks later, the Twins' sister was picking up the daughter of one of the Twins from day care when they were both kidnapped. The Twins had to pay a ransom of $250,000 to get them returned. I started to rethink how I'd been protecting my family from the people I dealt with. I stopped bringing any of my crew and associates around my family.

T had been out for a year now and we didn't talk as much as we used to. At times I felt he was a little jealous. He tried to hate on me with a couple of my clients and take them away from me. He would tell them I was his little brother and that buying from him was the same as buying from me. It made me sick to think about how proud I'd once been to have him call me brother.

He even went and told the Twins, "Jeff must be hot or working for the police, because how could he have gone so long without catching a case?"

Even though I was mad as fuck at T for saying that about me in the streets, I didn't have enough heart to confront him.

Although I didn't drink, smoke, or take drugs, I had developed one bad habit over the last couple of years: gambling. Weekend nights was craps time in the parking lot behind the Jack in the Box on Euclid and Federal. It was me, Hump, Black Blood, Tank Bo Bo, and this one professional gambler named Herbert. We were all huddled in a circle shooting dice for $500 a hand.

I was breaking motherfuckers left and right, rolling sevens after sevens after sevens. Then somehow Herbert slid in some trick dice and broke me for $60,000. I kept going home, getting ten thousand at a time, and Herbert kept taking it from me. I was, like, *How's this fool keep breaking me?* I was naïve to trick dice.

I should have never gotten into gambling. It just wasn't my forte, even though we did it all the time. The Twins taught me how to gamble on the many trips I took with them to Las Vegas. They were huge at Caesars. We hustlers would come out there in droves on private planes or in caravans. Caesars loved to profile our cars out front. They gave us everything we wanted—presidential suites, Jacuzzis, limousine service. The Twins would bring eighty Gs for the trip, Caesars would break them, and they'd be right back on that private jet to L.A. to pick up another sack of money and bring it back to Vegas. I'd blow thirty Gs over a weekend.

I took Carmen and other females from time to time, but Vegas being a players' playground, we tried to have a code that you didn't bring sand to the beach. Vegas had hookers from every country in the world. I fucked off a lot of money there, especially at the prizefights, but I never felt comfortable. I always felt I was being watched. I remember the Twins once got pulled into the police station and shown pictures of themselves in Vegas rolling dice. So I knew the Feds were hot on our tails.

Another spot we frequented several times a year was Montego Bay, Jamaica. It was my treat to the crew for doing good business. That was a time for R&R—no stress, no cell phones, no pagers. I bought first-class round-trip tickets on Air Jamaica for us, and we checked into the Wyndam Rose Hall, a plush resort on the beach. It was just like the places I saw on TV as a kid and dreamed about visiting someday. We hit the Royal Palms dance club where the local island girls danced in the nude and fucked to make a living.

Most successful dealers don't do business during the Christmas holiday season. It's bad luck. We usually shut down operations, enjoyed our families, spent money, and had a good time. I would spend fifteen, twenty thousand on shopping sprees for my mom, my sister, my niece, my son, and his mother. Even though it was over romantically between Tammy and me, we still had a cool relationship because of our son.

For Christmas '87, I put a down payment on a new I-Rock Z28 Camaro for my mom along with a mink coat and a trip to Hawaii for her and her best friend. I bought Carmen an $8,000 promise ring with matching baguette diamond earrings and gave her a couple thousand to buy her family some gifts.

I told my clients I was going to shut down for two weeks. It put a hurt on them because the lower-level dealers always had to keep slinging. One up-and-coming dealer named Sweet, who lived in a neighborhood on the South Side, started calling me and calling me saying, "Jeff, man, I gotta get some work. It's on."

I felt a little sorry for the youngster, wanted him to have a good Christmas with his family, so I decided to break the rules this one time.

I had about half a key left that was soft. So I said, "Let me go ahead and cook this half a key up for you and then that's all I'm doing until after New Year's."

I told Carmen to go ahead and rock it up for me. Not a half hour later, I was on the phone in the living room when Carmen calls me into the kitchen and says it's not cooking up right. Instead of forming a plate, the cocaine gel was beading up into little balls. I cooled it down and tried to melt it again, but it still looked like soap powder someone had spilled water on.

I hadn't gotten that batch from the Twins, but from another connection I used as a backup when the Twins were out. Big Bird was a rich L.A. baller. When I got him on the phone and explained the problem with his shit, he said, "That's cool, Jeff. Bring it back to me, make sure it's the same weight, and I'll give you another half key or your money back."

I said, "Man, I can't come all the way to L.A. You gotta come here."

"I ain't doin' nothing, man. It's the holidays. You know we don't roll like that."

I said, "Fuck it. Where you wanna meet at?"

I called up Sweet and told him we were going to make an L.A. run. Since I had to go up there, I wanted to drop off my '64 Chevy in L.A. for some repairs. So Sweet followed behind in Carmen's Sentra with the bad drugs and about $50,000 under the spare tire.

"When you go through border," I told him, "just be normal and smile."

In Orange County, sixty miles north of Mexico is the San Onofre border checkpoint. It's where they check cars heading north and south for drugs and aliens in case they missed them at the actual border.

As I pulled up to the checkpoint, I was on the phone and not paying attention. The border agent flags me on through ahead of Sweet. A minute later, I look in the rearview mirror and I see this fool leaning down in the seat. He must have been budded out, smoking some weed. So as he approaches the checkpoint, I see they don't flag him through. They're having some dialogue back and forth. Next thing you know, they wave him over.

I immediately pull off to the shoulder and watch through the rearview. In minutes, they've got Sweet out of the car, the trunk of the Sentra up, and then the Foot Locker bag sitting on the roof of the car.

My stomach starts turning and knotting up. I punched the Chevy, went to the first exit, got off the freeway, turned around, and headed back toward San Diego. I saw a Denny's, pulled off the Freeway, dropped the car down to the ground, locked it up, put the alarm on, and found a taxi. I'd come back for the car when things cooled down.

I asked the driver, "How much to take me to San Diego?"

"A hundred."

"Bet," I said, and peeled him off $300. "Get me there fast."

As I passed through the border, I saw Sweet in handcuffs on the other side of the freeway, and the car was surrounded by federal agents.

Back in San Diego, I got Carmen on the phone, and I told her in a panic, "Baby, we got major problems. Sweet was busted at the border in your car. They gonna run the tags and it's gonna come back in your name. And there are bills in the glove box with my name and address on them."

So she starts freaking out.

I say, "Listen. All you have to do when they get at you is say, 'I loaned the car to my boyfriend.' That's it. You don't know nothing. You don't know Sweet. You don't know nothing. You got that?"

"Yeah, baby. I got that, I got that."

Then I told her to get everything out of the apartment—all the glass pots, all the baking soda, the triple beam scale, all the Ziplocs, everything. "Scrub the stove down, clean everything, until there's no residue, no hint of cocaine anywhere in that house." I said she should get a motel room because I was scared to go to her apartment.

That night, we drove to a Wendy's and tossed everything in a Dumpster. I left Carmen in a motel room down in Mission Valley and went around to all of my stash houses, collecting all of my money and taking everything out of my mom's house. I told Tammy what had happened and she put my money in a safe I kept in her mother's attic.

Back at the hotel we were hiding out at, Carmen and me just lay in the bed staring at the ceiling all night, discussing the situation and all that could go wrong. We were lying with our shoulders touching, but every minute seemed to push us apart. She was still wearing that promise ring, but everything had changed.

Sweet was booked into the federal jail and he started making calls. First thing he did was to call his girlfriend, who was six or seven months pregnant at the time. They called me up on the three-way but I didn't want to be taking any calls from the Federal Detention Center because I knew they'd be taping it.

He said, "Jeff, man, I got busted. I'm downtown. I need to get out of this place."

"What's your bail?"

"I don't have bail yet. I have an arraignment Monday morning."

I said, "Okay, I'm gonna do what I can. I'm-a get a lawyer for you. We have to get off the phone now. Have your girlfriend give me a call."

She told me what Sweet had told her: that they found the money and drugs and were asking him to cooperate. He'd supposedly told them no. I asked about the stuff in the glove box. She said they took it. So they knew who I was. They knew who owned the vehicle, and they possibly knew who the drugs belonged to.

Next morning I called Hump and Jake, who knew one of the best criminal defense attorneys in the city.

We set up a meeting and the first thing this motherfucker says is he needs money to start the case. He wanted $25,000 up front and another $35,000 later. I said I'd give him his $25,000 right now and then hit the streets to raise the rest of the money so Sweet could post bail.

I didn't want to give up too much money for some nigga I hardly knew. Plus, there was always the chance Sweet would snitch me out no matter how I did for him. If I did end up going to prison, I wanted to have as much loot as possible stashed away.

The lawyer went down to visit Sweet and find out his state of mind. Our goal was to get him to take a deal right away, ask for a speedy trial so he could do the time for the case, and then we'd take care of him. At least that's how I thought it was going to go down.

After two or three visits, the lawyer informed me that Sweet was getting more and more nervous and fidgety, and he wanted out of jail. I had problems raising the bail money. It was only $15,000, but I'd already invested $25,000 in legal fees and I wasn't going to put up everything I had for that dude. Sweet's homies and family members didn't give a fuck about his black ass. And I was worried about communicating with them because I knew the Feds were pushing hard to break Sweet. He started calling me more and more, talking crazy: "Get me the fuck out of here! I can't take this shit no more!"

I knew this nigga was going to flip. The lawyer even told me if we didn't get him out soon he'd bitch-up and make a deal. After a month, the calls from Sweet stopped coming, and the lawyer got a letter saying he was being relieved of his representation of Sweet; the U.S. Attorney had gotten him to accept a court-appointed lawyer.

Me and Carmen went to the lawyer's office for a meeting. It was strange to have Carmen there because I was starting to feel I couldn't trust her. People were getting in her ear, telling her that I'd probably walk and she'd end up doing my time because the drugs and money were in her car. We were getting suspicious of each other and we began to argue constantly.

The lawyer proceeded to tell me exactly what I didn't want to hear: that at some point in time the Feds were going to come for me. He said it would happen within a year because that was the statute of limitations for drug offenses.

After the contact broke off between Sweet and me, I consulted with the Twins, who'd heard about my problems through word on the street. I explained what had gone down. My money was kind of short because my operation had all but shut down and I asked them to hook me up on consignment to make back the cash and drugs I'd lost, plus the $25,000 attorney's fee. They could have just told me to fuck off because I was hot, but they were loyal.

I went on full-scale hustler mode, sleeping all over town, a different apartment every night, different hotels, always trying to make it harder for the Feds to keep me under surveillance. If I was going to the pen, I wanted to have some bank put away to make sure my family was taken care of. I started liquidating all my assets, selling off my cars. I sold the white Mercedes SEC to a guy from

Detroit and it broke my heart. It was my pride and joy, but I had to let it go.

Running around became a routine for me, and I soon got comfortable and careless again. There were times when I just forgot about all my problems. I was back to enjoying life again, even spent another week in Jamaica. When I returned to San Diego, my relationship with Carmen was even more strained. She lost her job as a meter maid because the Feds had questioned her at work about her relationship with me and about her vehicle. She was stressed, I was stressed. Our arguments became more frequent and more hostile.

It was July 1988 and my twenty-fourth birthday was approaching, so I told Carmen I wanted a hell of a party. My budget was $10,000. We hired Cool T, the hottest DJ in Diego. Carmen hooked up with her girls to form a planning committee—they called themselves the Ladies of the Eighties.

I had the Grand Ballroom at the Omni Hotel in Horton Plaza. Carmen bought me a huge birthday cake decorated with dice. It said: HAPPY 24TH MR. JEFF.

Carmen had invited all sorts of people I was cool with. Even several of my girls on the side were there, so I kept on the dance floor, trying to stay away from them.

Dana was there, too. Dana was an ex-friend of Carmen's from high school. They knew about each other, but Carmen didn't know it was getting serious between Dana and me. With everything seemingly out of control around me, Dana was a calming force. I'd even gotten her an apartment with the same furniture hookup I'd gotten for Carmen so that I'd have another place to lay my head and chill out. Dana worked at a hospital and came from a big family who lived around the corner from Carmen's parents' house.

The apartment I got for Dana was a place I could rest and relax. And it had a gas stove so I could cook my stuff. At first, I'd tried not to cook too much there, but it was the one place the Feds didn't know about, so I felt safe. Tank Bo Bo was the only one who knew about that spot.

As the party began to fill up, it seemed like half of San Diego was there, everyone dancing and having a good time. There was no drama. Toward the end of the night, I made an announcement. We were raffling off trips for two to Jamaica and Hawaii. My sister Cali Slim won the Hawaiian trip and someone else won the Jamaican one. I wanted to end the night with one great splurge of extravagance. I wanted that birthday to be the party people remembered me by, one that they'd remember for years to come.

I remember trying to figure my escape from the party, because all my girls were trying to guess who I'd leave with; there was no choice except Carmen. Despite the tension, she still had my heart. We left in a limo, grabbed our luggage, and set off for Hawaii.

When we returned, money was tight. Most of my customers were hesitant about dealing with me because they knew the Feds were on me. I watched every dollar, and I started to catch Carmen pinching from my stash.

I put my home up for sale with a real estate agent. The end of the year was coming, and I had yet to be arrested by the Feds. I kept telling myself that my problems would come to nothing, but I never really believed it.

Carmen and me had been fighting about the fact I'd been spending less time with her and acting funny. I was less and less comfortable going to Carmen's apartment because of the pressure the police were putting on her. Eventually, I started sleeping anywhere but her place, and I'd been gradually moving on and building a stronger relationship with Dana. I really liked her because she never put pressure on me, never questioned me, and was low maintenance. She only asked for lunch money while Carmen's spending was really becoming a problem.

It was a week before Christmas of 1988, and I was kicking it at my house with a chick from the neighborhood—no one special— when someone started banging on the sliding glass door outside of my bedroom.

I got nervous, grabbed my nine millimeter, and slid the magazine in. (I always kept the magazine out because my son once found it and came in the room pointing it at me.) I told the chick to crawl on the floor to the room I kept for when my son visited. I crawled to the bathroom and stood up on my tub to peek out the window.

It was Carmen. She started screaming, "Open the door, Jeffrey! I know you have some bitch in there!"

I got angry and told her to leave or I was calling the police. That only pissed her off more. She started pulling up all of the fucking plants in my yard and throwing them against the house.

I told the chick to just chill, lie on my son's bed, and I'd get rid of Carmen.

Eventually, Carmen got worn out and left. I called her the next morning and cursed her out about coming to my crib tripping and told her she had to pay for my plants. She came by and started up again. This time she rushed me through the front door. That's when things got ugly.

Carmen had very long fingernails. As she approached me, she clawed at my face with them. The bitch knew about my bad eye and she was going for my good eye anyway. I realized she was just crazy enough to want to fuck me up for life, so I reacted by slapping her hard in the face.

She fell backward, lost her footing, and tumbled down my concrete steps. She scraped the stucco wall as she was falling and left a streak of blood across it.

I felt awful. I'd never put my hands on a woman before. Carmen jumped right up, bleeding, and charged me swinging both fists. I put mine up, and we were going at it in my front yard. We stood there blasting on each other back and forth. I never put all my strength into my swings, but Carmen threw hers as hard as she could.

Finally, the old black lady next door came out and made us stop.

After the fight, Carmen left—but not without knocking the mirrors off my car.

A few mornings later, on December 21, I was over at a different girl's house when my cell phone went off. It was the real estate agent telling me he had a buyer and that I needed to come to my house right away. At eight in the morning, I jumped in my '86 Chevy Celebrity and headed for my home.

As I climbed the hill and made the left turn onto my street, I noticed some cars I'd never seen before and a couple of vans, but didn't think much of it. I called Al, told him I was pulling up. He was out on my balcony, and I saw my wrought iron gate was open. Something didn't seem right, so I didn't pull over, just kept rolling at two miles an hour.

Suddenly, federal agents jumped out from across the street, from the side of the house, from the neighbors' yard, from out of the vans—seven or eight of them, drawing their guns on me.

"Freeze! Stop!"

When I saw them coming, I stepped on the accelerator a little bit, but decided not to take the chance. Those motherfuckers would have killed me.

"Hands on your head!"

One of the agents approached my car, opened the door, and had me drop to my knees and lie on the pavement.

I was scared to death, but I also had a deep feeling of relief that it was finally over. I didn't have to run anymore, didn't have to hide anymore.

I was thinking, *Whatever's going to happen, let's do it. I gotta put this behind me and get on with my fucking life.*

I'm on the ground, then one of the agents grabs my cuffs and pulls me to my feet. I later learned he was Fred Matthews, the lead Fed on the case, who was spearheading the San Diego Drug Task Force—a multiagency unit composed of agents from the DEA, ATF, IRS, and the local San Diego PD.

Matthews walks me into my house. My car's still running in the street. I see my neighbors looking out their windows in disbelief and shame. I had told them that my dad owned a dealership in L.A. to justify all my cars.

They'd ransacked everything. The cushions off my sofa were on the floor, all my cabinets were open, things were dumped over and turned upside down.

Sitting me on a chair, they began to read me my rights, and my cell started ringing. One agent grabbed the phone and answered it.

"Hello?" he was saying. "Do you need to buy something?" I could tell that whoever was on phone was saying, "Who is this?"

The cop just stuck to his lame game, asking, "You need to buy something?" My client hung up on him.

Then they asked me if I wanted to cooperate.

I told Matthews, "If I'm under arrest, take me to jail."

He said, "I can give you an opportunity to help yourself."

I told him again just as calmly, "If I'm under arrest, take me to jail." Then I gave him the statement: "If you can't do the time, don't do the crime."

He continued to try and engage in conversation about other dealers and about my house. From the corner of my eye, I saw the real estate agent sitting there on my couch. He looked at me, shook his head, and shrugged his shoulders, as if to say, "I'm sorry."

Once they realized I wasn't talking, they secured the house, locked the door, and put me in the back of a van. As I looked out the window at my home, I knew it was the last time I would ever see it.

As we cruised down the MLK Freeway to the Federal Detention Center in downtown San Diego, the officers never spoke to me again. Outside the Detention Center, an officer reached up and hit a button. A big chain gate rose to the ceiling and we went into a tunnel that led underground. They got me out of the van and walked me through several electronic doors to the booking tank. They stripped me and gave me my first cavity search. I was made to raise each foot behind me and wriggle my toes, spread my buttocks, and lift my testicles. It was the most humiliating experience of my life. I felt all these strangers watching me. I had never felt so vulnerable and helpless.

They put me in an orange jumpsuit with BOP stenciled on the back: Bureau of Prisons.

I was fingerprinted, photographed, and asked one more time if I was interested in helping myself.

I said no.

SIX

SOUL ON ICE

After the booking I was moved into a holding cell with several other inmates. I paced the floor, kept my mouth shut, and waited to be transferred to the inmate receiving area on the fourth floor. About three hours later, the federal marshals handed me over to two BOP guards who escorted me to receiving, where I met my caseworker for the first time.

Ms. Wilson was an older black woman from southeast San Diego. As soon as I sat down in front of her, she started asking questions: Did I have any enemies? Did I want to be in protective custody (PC) or in general population? Had I ever had tuberculosis? VD?

I told her that I was healthy and that I wanted to be in general population.

"I don't have any enemies," I said. "Not that I know of."

The cell I was issued was in B Unit of the receiving area. Each floor in the Detention Center consisted of four units—A, B, C, and D—each of which contained ten two-man cells. Every floor had a dining room, TV room, weight room, pool table, and a small kitchen.

My cell had a bunk bed, a stainless steel sink and toilet, and a one-by-two-foot window facing the west side of the city, where I could see the ocean and the aircraft carriers around North Island. There was a small desk, and a light was embedded in the wall for reading and writing. The floor was hard tile and the bare walls were fortified concrete—your voice bounced off them when you spoke.

My cellie was a guy named James. He was a cool brother but I could tell he was a smoker when he was on the streets. The dude stank like he hadn't bathed in days. He'd caught a small case, was looking at four to five years. I threw my bedroll on the top bunk, got myself situated, and me and James began to size each other up.

I said, "How long have you been here?"

"I've been here for a couple of weeks, man."

"Where you from?"

"I'm originally from L.A.," he said, "but I caught a case out here in a sting operation. What're you here for?"

"They got me on some conspiracy charges. They caught this other dude with some money and cocaine, and they're trying to finger me for it. Just fucking raided me at my house this morning."

"That right?" James said. "You ever done time before?"

"Nah, I never did no time before." The next thing I asked him was, "When do we eat around here? What's the food like?"

"Man, the food is pretty good. I can't complain. Sometimes we have T-bone steaks. They got burgers, baked chicken . . . Man, you never heard about the Feds?"

I told him I'd only heard about state prisons.

"The Feds take care of you up in here, man—air conditioning, heaters, library, barbershops, good food. It's incredible. The only thing about the Fed system," James told me, "is you don't get conjugal visits."

With all the drama I'd been through that day, getting laid was the last thing on my mind, but as soon as James mentioned it, the reality of my situation began to hit me.

I said, "You can't get no pussy?"

"Only way you get that is if you can get lucky in the visiting room with your old lady. It can be done."

At ten o'clock that night a guard came to our cell and said, "Count time," and I'm, like, "Again? Where the hell they think we're going to go in this high-rise?"

"They count us four times a day," James said. "You got a ten o'clock count at night, you got a 4:00 A.M. count, another ten o'clock in the morning, and another at four in the afternoon. The most important count of the day is the 4:00 P.M. count because they call that one in to Washington, D.C." All inmates are released in the morning or early afternoon, so by 4:00 P.M. the federal system has an actual body count, which they report to Washington. "If you're not where you're supposed to be for that count and it comes up short, they shut this joint down, find you, and lock you up in the hole."

Disruptive inmates also went into solitary confinement, or "the hole," James told me. The hole was located in the basement of the detention center, the most confined and secure part of the building. If you went to the hole, you had to deal with a guard called Big Bubba, who had been at the center since it opened back in 1975. Inmates said he was the only guard who never used handcuffs, that he was so large and intimidating he could just grab your arm, say, "Let's go," and no one would dare challenge him.

After the 10:00 P.M. count, the guards opened up all the units for the inmates to get out for an hour and a half before lockdown.

I went straight for the telephone to call my mother and Carmen and tell them I was okay.

When my mother picked up the phone, I said, "Hey, Moms, it's me," and she started to cry. "I'm okay," I told her. "Moms, I'm okay." But she kept weeping in my ear.

"Is anybody bothering you?" she said. "Don't let anyone bother you." She kept saying that and kept crying and it pissed me off because I was trying to be strong. Within a few seconds, though, I just broke down crying and hid my face in the corner.

I got myself together because the phones automatically cut off after fifteen minutes and there was still business to take care of.

"Moms, make sure you get all my money from Dana," I said. "Get all my money and stuff from Tank Bo Bo, and make sure everything is safe. I can't talk too much on this telephone because they listen in. And don't worry about putting no money on the books for me, because I had $1,500 in my pocket when they arrested me, so I should be okay." Even though I still thought I could make bail in a matter of days, I wanted to calm my mother down as much as possible, and not trouble her with having to put money on my detention account for basics like toiletries, sodas, and snacks.

"Anyway, Moms, I gotta get off the phone now but I'm-a call you first thing in the morning."

"Okay, baby. Momma's here for you. I'm praying for you. I love you."

Carmen never answered her phone. I was pretty upset because I'd expected her to be waiting for my call. Plus, I was kind of worried about the Feds fucking with her, getting into her head, trying to get her to flip on me.

Dana picked up after the second ring. She was so happy to hear my voice and so concerned about me.

"Everything's okay," I told her. "Just help my mom. You know, be there for her right now because she really needs some help."

To take my mind off my family I decided to go and chill in the TV room. All the blacks were on one side of the room watching Arsenio Hall, and all the whites and Mexicans were watching soccer on another TV. Officer Allen—a brother who would eventually become a close confidant—yelled, "Lock down!" at 11:30 and all the inmates slowly started walking back to their cells.

Back in our bunks, James continued giving me the rundown on how things worked in detention, schooling me on how to do my time. We talked for a long time about our worries and our fears. In my mind, I still thought it might be possible that I'd walk out within forty-eight hours, hug my mom, see my son, see Dana and get some loving.

When I finally fell asleep around 2:30 in the morning, it seemed like only seconds had passed when the graveyard officer woke me up for the 4:00 A.M. count, shining a flashlight beam in my eyes.

I never got back to sleep that night. I wished I could turn back time, keep my job at the printing company selling my little dimes of weed. I wish I had told T that I wasn't interested in hearing about the new drug that was going to make us all rich. That I never sold crack, never got past being a petty hustler. But I wanted so much to be like T, I wanted so much to be the man. And I had become like T—only I was *bigger* than T. I *was* the man. And this was the price that came with the game.

At that moment all I wanted was to see the streets again, spend the holidays with my grandparents, and watch my son grow up.

At 6:00 A.M. the guard quietly opened the doors, and whispered, "Chow time." I was hungry to the point of nausea by then because I hadn't eaten since the night before my arrest. Even when I'd been poor, I had never gone that long without a meal.

Two inmates were serving food on the breakfast line. One of them was a burly white boy with a fucking swastika right on his neck, so I made sure I had my stone look my face. I didn't want to appear soft or nervous. I picked up French toast and oatmeal with half an orange, a banana, and a half-pint of milk and sat at the table closest to the elevator, where all the brothers ate. I stayed quiet and observed.

A lot of inmates went back to their bunks after breakfast but I stayed and hung around the common area, so I could be first to see Ms. Wilson and fill out my application for visiting rights. The longer I went without hearing from anyone, the more my confidence in getting out within forty-eight hours slipped away from me. Ms. Wilson told me that I couldn't even fill out the paperwork until they moved me to the twenty-fourth floor, which is where they kept you while you fought your case. I spent the rest of the day telling myself to toughen up, that I couldn't be emotional, couldn't be seen out on the floor whining on the phone with Moms. They transferred me upstairs the next day.

It was just like being back in my neighborhood.

There was Schoolyard, Cutty Shark, and Smokey—three niggas from the Syndo Mob—Dave and Dory from the East Diego crew, and my boy Jake from Skyline. He came up on me almost as soon as I got out of the elevator.

"What up, Hard Head?" he said. "We heard you got knocked. We were wondering how long it would take you to get up here. Thought maybe you PCed-up." (He meant going into protective custody.)

I said, "I don't have no reason to PC-up. I'm in here like everybody else."

A bunch of the other homies came around saying what's up and immediately they started asking, "What did they get you for? Whose case are you on?"

"I'm not on nobody's case. I got my own. You all know that

young nigga Sweet from the South Side who got cracked last year at the border? Well, he finally flipped on me. So they just raided me a couple days ago and hit me with two counts: conspiracy and possession of a controlled substance with intent to distribute. I'm trying to get bail right now."

Jake told me he had an empty bunk in his cell. "I'm-a put you in there with me," he said, and took my bedroll. I followed behind him carrying my toothpaste, hair grease and other cosmetics, and the snacks James had given me.

By the morning of my bail hearing a week later, I'd gotten pretty scrubby because I didn't know how to shave with a razor—I'd always used clippers. I was skinny, my head was nappy, and I was looking crazy. I felt embarrassed. I knew my family would be in court and I didn't want them to see me this way. I also didn't want to look at them and get emotional at the hearing.

Flanked by two marshals, I walked into court. Mom, Dana, and Tammy all broke down in tears. Even though Tammy and me weren't together anymore, we had stayed close, and I guess she felt obligated to see what fate had in store for her son's father.

When I met my court-appointed attorney, Virginia Eli, I thought, *They're gonna hang me! I can't have no black attorney!* Everyone wanted a Jewish lawyer because they were believed to have the best connections and be the best at convincing an all-white jury that their client is a victim of circumstance.

It was the first time I saw the prosecutor, U.S. Attorney Amelia Meza, and I would never forget her. She hated me with a passion, even though she didn't know me. She seemed to hate me through paperwork, from the statements Sweet and the agents gave her, and her belief that I was a criminal.

I was trying to look humble, like I hadn't done anything wrong, when the bailiff announced, "The court will now rise. The Honorable Judge Gordon Thompson presiding."

A tall, burly white man with a receding hairline in his late fifties wearing black-framed schoolboy glasses took a seat at the bench. He didn't look nice. He looked like maybe I stole the battery out of his car one day and now this was his chance to get back at me.

After a little courtroom chatter, Judge Thompson set bail at $300,000.

"Objection!" U.S. Attorney Meza said. "Your Honor, we believe that Mr. Henderson is a flight risk. We believe bail should be denied based on the following facts. . . ."

To me it was all just *blah-blah-blah*. It was bullshit, calling me a flight risk. They just wanted to keep me locked up because Sweet was going to be on the streets soon and they didn't want me to be in contact with him. The judge overruled the objection and my lawyer whispered, "I'll be by to see you in the next couple of days."

When I turned back to my mother and family, I hoped that the look on my face told them I loved them.

It was lunchtime when the marshals brought me back to the holding tank, and a brown paper bag was waiting for me: bologna on white bread, an apple, and orange juice. I spent the rest of the day in that tank, waiting for the other inmates who'd had hearings so we could all be transferred back together.

It was a long day. I started pacing the floor and having self-talk. I started praying to God. Now I wanted to be religious. I always did love the Lord, but I guess I forgot about him when I was out there selling drugs. I started singing this hymn I'd learned as a young kid in Bible study: *Jesus, I know you gonna make a way for me. Oh, Jesus, I know you gonna make a way for me.* I must have sung that a thousand times in that holding cell and walked fifty miles.

When I got back to the unit, all the homies were slapping dominoes on the table. I hate fucking dominoes. I hate the noise. I didn't even know how to play the damn game. They didn't even put a blanket down to slam them bones. I just kept hearing *CRACK! CRACK! CRACK!*

How in the fuck are these guys sitting here playing games? And we're all getting ready to go to the penitentiary? We're fighting for our lives! And these other fools were shooting pool right in the middle of the fucking common area. What the fuck do you need entertainment for when your life is on the line? Later, I'd learn that everybody does their time their own way.

All the homies stopped slapping dominoes and got quiet at the brothers' table when they saw me walk in from court.

Cutty Shark asked, "Jeff, how much is your bail?"

"Man, they hit me with three hundred thousand. I don't know if I'm gonna be able to make that, but I'm sure gonna try. I gotta talk to my family."

"Was that fool Sweet up there?"

"Nah, I didn't see him. They probably won't bring him in unless I go to trial. Listen, I need to get on the phone now."

I had to line up some family and friends to put up their property for my bail. I was telling myself, "Now we'll see who's down with me, who's got my back."

When I called my pops on the three-way with my mother, he was in shock. He didn't know it was that serious. Hell, I didn't know it was that serious.

I was, like, "Pops, I need houses. I can't put up no cash because the Feds'll want to know where it came from. Houses are the only thing they're gonna take to get me up out of here. You think Auntie Barbara would put her house up?"

"It doesn't look good, son," he told me. "Aunt Barbara's house is community property and your uncle Harold said 'not going for it.' I only have a little equity in my house, and Grandma and Granddaddy don't have much in theirs, either."

"What about Auntie Joyce in Chicago?"

"It doesn't look good, son."

"Damn. What we gotta do, we gotta meet with the lawyer and see if she can get the bail dropped down to like a hundred thousand. If we could do that, there'd be enough equity in your house and Grandma's house to get me out."

Since Dad was in L.A., I asked my mom to get in touch with my lawyer to talk about getting the bail reduced before my upcoming evidentiary hearing.

"Baby," she said, "you know your momma don't know about that stuff. You gotta talk to your sister or Dana. Momma don't know those things."

"You've got to start learning, Moms. This is my life. You've got to be strong. Get Dana on the phone and I'll call first thing in the morning. Moms, you gotta start learning this. Don't be intimidated."

Next I called Dana, who'd already heard from the Twins. They said they were going to hire a lawyer for me. But I'd end up spending weeks trying to get ahold of them, as they never answered their phones. I figured they were scared I'd turn on them, even though I could have done that upon my arrest and never served a day. That's friends for you.

Two days later, Officer Allen was back: "Roll 'em up, Henderson. You have an attorney visit."

I jumped up, put Vaseline on my face—it was raw from my first razor shave—and Officer Allen took me to the visiting room. Ms. Eli was at a table in the back of the room.

"How you doing, ma'am?" I said. "I couldn't wait to see you."

We discussed bail for a few minutes when she cut me off, saying, "I'm going to be up front with you, Mr. Henderson. These charges are very serious. You have no chance of getting out unless you can raise property."

"What are the chances we can get it reduced?"

"It would be very difficult. All the arresting officers were there. For some reason, the U.S. attorney doesn't want to see you on the streets."

"Damn," I said. "I didn't kill no one. I wasn't even a banger."

"You were a major dealer, Mr. Henderson. They've had a line on you for the last couple of years, even before the Sweet case."

I told her I still wanted her to try to get my bail lowered. And she did try, but then the judge revoked it completely. Sweet pleaded to eighteen months, had done a year already, and was heading to a halfway house to do the last six months. I was especially worried because the government hadn't even revealed its evidence yet. I didn't know how much they had against me.

Back in the holding tank, I was talking to God again: Why me? I know I did wrong, but I helped people, too. I took care of everyone I knew. Bought cars for people, fed people who didn't have anything, and always donated cash to the Skyline Youth Center. I never intentionally hurt nobody.

I blamed everyone but myself.

Jesus, I know you gonna make a way for me . . .

It had been a big court day for a lot of homies in my unit. Schoolyard and Cutty Shark had just come back. Everyone was stressed and no one was talking. No dominoes, no pool. Niggas were lined up at the phones trying to reach out to the world.

Everyone was mad as a motherfucker.

Things were still tense the next morning and everyone was still trying to get their calls out. Schoolyard came out of the weight room and started pacing back and forth behind me while I was on the phone with Dana. I'd heard he was on edge because he was having trouble with his woman—we all were.

Earlier that morning when I tried to call Carmen, she still wasn't home. I'd been hearing a lot of talk about what she'd been up to at the clubs. I knew I had to let her go and put all my chips with Dana, but I was pissed off anyway.

"Nigga!" Yard screamed in my ear. "Gimme that motherfucking phone!"

"Baby, hold on," I told Dana. To Yard, I shouted, "Nigga, can't you see I'm on the telephone?"

Turning away from him, I could see his reflection in the window in front of me. When I saw him raise a fist, I ducked. He missed me and I jumped up. We started getting down like we were back on the streets. I didn't want any problems, but I couldn't be a punk. While we traded blows, I was surprised that he wasn't really putting them on me since his rep on the streets had him as a hell of a gangster. I hadn't been in a squabble in a long time, so it made me more confident.

After a few seconds, someone screamed, "Guard!" and we went back to our cells without either of us having really connected. When the guard went away, Yard came right in my face, saying, "Nigga, you want more?"

I said, "I got no beef with you, Yard. All you had to do was say, 'Man, can I get the phone?' and I'd have let you use it." It was squashed after that.

Word had hit the street about my fight with Schoolyard when I finally got Carmen on the phone. First thing she said, "Baby, what happened? Heard you had a fight with Yard."

"We had a little scuffle over the phone. He was having trouble with his woman. And by the way, where the fuck you been at? I try to call you in the morning, at night, and you're never there."

"You know I gotta work, and I still have a life."

"What about me? You say you love me and I can't get you on the telephone?"

"I don't want to upset you. I gotta tell you something. One of the Twins got killed."

I was in shock. "What? How?"

"They're saying he was kidnapped and held for ransom." Carmen told me they beat him while he was on the phone with his mother. When his people couldn't come up with the ransom, they shot him and left him in a Dumpster in Long Beach.

"Do they know who did it?"

"No, but it was probably someone close to him. They're saying it might've been the same people who kidnapped his sister and daughter last year."

I went back to my cell. I'd been with the Twins for years. I never got to say good-bye and now I'd never see him again. I thought about the week we'd spent in Maui together with our girls, about how much he liked riding those little pedicabs and swimming in the ocean. Then I thought about the fact that he would still be alive like me if he'd gotten busted.

My first visiting day came two weeks later. I'd put Dana on my visitor list, not Carmen, but she showed up anyway.

"You ain't on the list," Dana told Carmen.

"Yes, I am," Carmen said.

"We gonna see who goes up to see Jeff." When the guard checked the computer and told Carmen she couldn't see me, Dana started laughing and clowning on her.

I was groomed, had gotten my shaving down to a science. I had on clean khaki pants and a clean khaki V-neck. The homies had told me what you could get away with in the visiting room. There was a peephole for the guard but you could see his feet under the door whenever he was watching you. When the guard wasn't around, they had a camera, but it usually wasn't working. It was a gamble.

Dana was wearing a beautiful baby-doll dress when I saw her in the room. I kissed her, we sat down at a table, and I started touching her up and discovered that she wasn't wearing panties.

I was, like, "Damn, baby, that's what Daddy needs. Slide back a little bit."

There were some families there, so I was nervous, but stronger emotions were pushing me.

I whispered, "Baby, put your hand in my pocket. I cut a hole in it, Boo."

After three or four visits, we had it down tight. Every visit was like a dream, it didn't seem real—to be touched, to feel her against me, to smell her skin. For the first time in my life I wanted to cuddle and be all lovey-dovey.

When three months had gone by, I'd had two or three more hearings and started getting with the program. Once I'd accepted the fact that I had to do some time, I slept better. And I had things hooked up with a cool officer.

Whenever he'd see Dana, he'd tell me, "You've got a fine-looking woman."

"She's got some fine-looking friends, too," I said.

"Oh, yeah?"

"Yeah, man. Look, I'm-a have her hook you up on the outside."

Before long, he was dating Dana's friend Demita. At least once a week, Demita would bring him two lunches in one bag and he would leave one for me in the mop room when I cleaned the unit after the 10:00 P.M. count. My favorite smuggled-in meal was a carne asada burrito with beans, cheese, sour cream, and guacamole from Roberto's Taco Shop. I ate every one of those meals like it was my last because there was always the chance my officer friend could get busted. I didn't even brush my teeth after those treats because I wanted to savor the flavor all through the night.

My court dates started heating up. At my first evidentiary hearing, they said they had half a key of that fucked-up dope with my prints on it, $37,000 in cash, and a phone book full of calculations for drug sales in my handwriting.

Ms. Eli leaned over. "It doesn't look good."

"What do you mean? I didn't get caught with anything."

"With conspiracy, you don't have to. Sweet says the cocaine was yours, you say it wasn't. They say the prints are yours, you say they're not. It does not look good, Mr. Henderson."

"What am I looking at?"

"The minimum mandatory is ten years to life."

My heart stopped. "*Life?* Who did I kill? This doesn't make no sense!"

Back in the unit, everyone rushed me wanting to know what had happened. It was always like that, every man wanting to know about the next man, when they should have been worrying about themselves.

Two days pass and a guard tells me, "Roll 'em up, Henderson. You have a court date."

"But I wasn't supposed to go back for three weeks."

The guard took me down to the basement and under the street to the courthouse, but he took me up in a different elevator this time.

I found myself in the U.S. Attorney's Office. For some reason they were being nice to me. They even offered me food.

Oh, I thought, *shit.*

Next thing, an agent named Ashcraft and a few other officers, including one I used to see on the streets called Big Red, sat me down at the table and asked me if I wanted to cooperate with them against my L.A. connection.

"I have nothing to say," I told them. "The drugs were Sweet's. I loaned him my car. I don't know anything else about it."

The guards came back and got me.

All the homies in my unit were pissed because some dominoes and pool balls were missing. The night before, I'd grabbed them when I was cleaning and put them in the trash. You can't play dominoes if you're missing three of them. You can't play pool if you're missing the eight ball. The brothers were giving the white boys and Mexicans hell about it, but at least I didn't have to hear all that noise anymore.

Around that time, I moved in with an older brother named Carlos from L.A. Jake was talking or snoring all the time and I needed quiet time to think and relax. Carlos was a Sunni Muslim, a very conscious brother, who always called me "youngster." He'd been a pimp and a heroin addict, had marks all up his arms and neck, but he kept his cell immaculate. He was in on a violation for a dirty drug test.

Since Carlos had more than twenty years' experience in the system, he started schooling me on how to do my bit if I had to do some penitentiary time. I would buy him food at the commissary on the roof by the basketball court. Once a week, we stocked up on avocados, nachos, Pepsi, Ramen noodles, and candy bars— Carlos loved the candy bars because the sugar rush helped calm his jones. When I bought him food, he shared more and more knowledge about life and survival in the penitentiary. Carlos told me I'd make it if I minded my own business, didn't gamble, didn't borrow, and kept away from the gumps—what we called gay guys in prison.

He even taught me the recipe for my jailhouse cologne: Take half a bottle of baby oil, a third lotion, add a couple cologne samples that we smuggled in, and use it to make you smell like a civilian for your visits. We called it Smell Good.

He also told me if my woman ever ran off on me, I could get someone on the outside to run personal ads in the papers: "Single black male. Incarcerated. Intelligent with a delayed future. Looking for single or married woman, looking for companionship."

He said, "You wouldn't believe how many women respond."

By August, I'd been locked up for eight months and the government offered to let me plead to 144 months. I rejected the deal for a jury trial and Ms. Eli laid my cards on table.

"Jeff, you can take a deal and be home in ten or twelve years. If you lose at trial, you won't be so lucky." I asked for her advice, but never got a straight answer. "You have to decide what you need to do for yourself. I can't make that decision."

She told me they would probably subpoena Carmen to testify against me. We had talked about getting married so that she wouldn't have to, but I had my mother give her money and move her out of town. The Feds raided Carmen's parents' house several times, but she wasn't there when they showed up.

Without Carmen, all they had was Sweet's word against me. It gave me hope. I got myself psyched that these white folks would let me walk out of jail.

Back at the unit, everyone was making deals. Yard got fourteen years, Cutty Shark got eight, Dave got ten, Dory got twelve, and Jake got three.

The first day of my trial, we had jury selection, and it made me think that there was a conspiracy between my lawyer and the prosecutor. Jury of my peers? There was not one black motherfucker in that whole jury box.

Ms. Eli kept asking me, "Are you sure this is what you want to do?"

I looked back at my family in the gallery and gave them a wink. In my blue slacks and white dress shirt, I was sure at least one juror would say, "He's just a baby!" and let me walk. I wanted to go home so badly that it clouded my judgment.

When I returned to my unit that evening, as usual, all the homies got in my face:

"Don't go to trial, man!"

"They gonna hang you!"

I was, like, "Man, they don't have shit on me. I never got caught with anything."

I just couldn't believe that I could be convicted without ever getting caught with drugs or money. I was in serious denial.

That night, I took a shower and turned in early. I lay in my top bunk, and stared through the cell window out at the ocean. When the officer asked me if I wanted to come out and eat, I told him I didn't feel like it.

At 2:00 A.M. I woke up to the sound of my cell door slamming open. The room flooded with flashlights and a goon squad of guards rushed in and told me to get out of my bed.

"What the hell is going on?" I asked as they cuffed me and walked me out of the cell barefoot and in my underwear.

"You know what's going on."

They took me down to the hole without another word. That morning, I was brought before the Detention Hearing Officer (DHO), who told me they were bringing attempted escape charges against me.

"Escape from where?" I said. "This is bullshit. I was on the twenty-fourth floor! Was I gonna fly?"

I spent another week and a half down there before I got to see Ms. Eli. She explained that bringing trumped-up escape charges was a typical bullshit Fed tactic. They were trying to build up their case against me, making up infractions they could bring up at trial. They took me back to the twenty-fourth floor four days before my trial, and the "attempted escape" stayed on my record for the rest of my years in prison.

The morning of my trial, I was taken to the receiving room to change out of my prison garb and into my slacks and shirt. As we walked back out, I saw none other than Sweet himself peeking through a window at me. He gave me a mean look, trying to stare me down.

I said, "Damn, what the fuck is he doing here?"

But I knew the answer, of course. He was there to testify.

At 9:00, the bailiff came around and told the marshals to bring me into court.

My whole family was there again, plus Dana, Carmen, and Tammy.

While the assistant prosecutor gave her opening statement, Ms. Eli told me to keep smiling and to maintain eye contact with the jury. It wasn't easy. I'd never heard anyone talk about me that way before, especially not in front of my family: I was called a thug, a criminal, a cancer on society, a dealer who sold poison in the streets.

When Ms. Eli gave her opening statement, she told the jury that I was a working man with strong ties in my community, that I was an upstanding citizen from a loving family. Jeffrey Henderson had nothing to do with Lamont Sweet or his drugs.

The testimony phase was a battle from the moment it started.

The lawyer I had originally hired for Sweet took the stand as a witness for the prosecution, but he'd barely started answering questions when Meza moved that she be allowed to treat him as a hostile witness. Judge Thompson granted her motion, but she still couldn't get him to give any damning testimony against me. She got frustrated quickly, dismissed the lawyer, and the judge instructed the jury to disregard his testimony.

The government then called Sweet to the stand. He appeared to be startled and ashamed, and he answered questions hesitantly. When the prosecutor asked him about his relationship with me, he said he'd worked for me as a dealer for three or four months.

I was looking him dead in his eyes, staring him down.

Sweet soon became hostile and uncooperative as well. The way I saw it, he was trying not to oversnitch. He was trying to say just enough to get by, and he was pissing off the prosecutor. She decided to end her examination of him. Still, he had done the job they wanted. I could see it on the faces of the jury, or at least I thought I did. On Sweet's face, I saw relief mixed with shame. That was the last time I ever saw him.

The next prosecution witness was the arresting agent, Officer Baylock. He was talking about the raid on my house when Ms. Eli, said, "Objection, Your Honor. This is absurd!"

Baylock had been trying to sway the jury by saying that my house "smelled like fresh leather" and that I had "more tennis shoes" than he'd ever seen. He was trying to make it seem like my lifestyle was so Gotti. Judge Thompson told the jury to disregard those statements.

When Ms. Eli called me to the stand, I was careful about my walk, made sure I wasn't strolling because I wanted the jury to see a humble, innocent guy. On the stand, I was very polite and respectful as I explained that I'd never even been at the border the day Sweet got busted.

Ms. Eli asked why I had made that *Baretta* statement about doing the crime and doing the time. I told her I was just being ma-

cho because I was upset at being harassed by the cops for crimes I
didn't know anything about.

The U.S. Attorney then began trying to cross me up, but her
game was weak. I looked her straight in the eye and at no point did
I feel that she had me on the ropes. She couldn't get anything out
of me, so she called me off the stand.

Closing arguments commenced a couple of days later. The
prosecutor explained why I should be convicted, removed from
society, and handed the maximum sentence. Her closing argument
didn't sound any different than her opening.

Ms. Eli told the jury that the government hadn't met its burden of proof. Lamont Sweet was a convicted felon and a street
thug who'd never held down an honest job. Lamont Sweet was an
admitted drug dealer; he was high at the time of his arrest. How
could anyone believe his testimony?

Judge Thompson read the jury instructions, and deliberations
began the next day. I spent the whole day in that holding tank, pacing and praying, asking God to forgive me.

Jesus, I know you gonna make a way for me . . .

Now I've become a religious fanatic. I'm a Holy Roller.

Still, a part of me knew I wouldn't get out of this. But when the
end of the day came and there was still no verdict, I saw it as a sign
of hope. That night I started telling myself again that I was going
to walk out.

"I'm gonna go home," I said. "I'm gonna see my family, make
love to Dana, and get me some Popeye's fried chicken!"

At one the next afternoon, I'd just finished my peanut butter
and jelly sandwich when the marshal came into the holding tank.
"Henderson, you ready? Verdict's in."

I was sweating beads. I started praying again and crossed my
fingers.

I hugged my lawyer for the first time. She said, "Jeffrey, I was praying for you last night. I hope we did it."

"Please rise," the bailiff said.

When Judge Thompson asked the jury foreman if they had reached a verdict, something strange happened, something I'd never seen throughout almost a year of court appearances. Two U.S. marshals came and stood right behind me at the defense table. This always happens when a verdict is read, but it was nothing I had ever experienced before.

The jury foreman I remember distinctly. He was a Filipino naval officer who always stared at me during the trial. With him being a minority, I thought he might feel a little compassion toward me.

"Guilty."

Each juror was then asked to give their verdict, and all the way down through to juror 11, it was the same thing.

"Guilty."

Before the last juror gave his verdict, I told myself, "I have one more chance."

"Guilty."

A rush of hopelessness ran through my body. I heard a scream. I looked back toward my family and saw Tammy collapse on the floor.

I just put my head down. Ms. Eli squeezed my shoulder and said, "It's going to be okay. We're going to appeal."

Everyone crowded around me at the unit that night, saying, "Big homie, it's gonna be all right. It's gonna be all right."

I was numb and shocked, when a few hours later, I was kicking it with one of the fellas in his cell and I suddenly broke down crying in front of him.

Dory told me, "Jeff, you're gonna be all right, man."

"I'm not going to be all right, man! I'm never gonna be all right! I can't believe this shit happened to me. I'm finished, man. I'm finished!"

I stared at my ceiling all night long. I was so tired, tired of the five-year run I had out there selling dope. The cars, the lying, the sneaking, manipulating women, all the wrong shit—I was tired of it. I just wanted to rest.

My family came to visit the next day. Moms started sobbing, then Dana, and then I did, too. Finally, I begged them, "Please stop crying! I've gotta suck it up. I have to do this."

I knew I had to be strong. I had to show all of them that I was strong, because I needed them to be strong for me as well.

At night in my cell after lights out, though, it was hard to keep up the strong front. I couldn't stop myself from thinking that my life was over. I prayed that I could take it all back: I never sold drugs, I never got involved with T and his crew, never even looked twice as my moms drove past him when she took me to live with her, and never even took that damn bike the cops had set up for a fool like me to take. I spent so much time thinking these things, going over every wrong thing I'd ever done, but every thought was laced with the fact that I knew I could never change what happened.

The emotions would just build until the point that I thought I couldn't take it. I knew I'd be locked up for a long time, and sometimes my mind would fuck with me, make me think that I would never get out. Lying on the top bunk with a view of the ocean, sometimes I couldn't even look out the window.

My sentencing came two months later. Judge Thompson took off his glasses and looked at me.

"Mr. Henderson, I'm bound by sentencing guidelines to give you the maximum sentence. I don't believe you deserve this sentence, but I have no choice. I'm bound under the law."

I took a breath.

"Two hundred and thirty-five months, to be served consecutively."

My head dropped to my hands. Nineteen and a half years. I thought to myself, My God. I'll be in my fourties when I get out. I'll never see my grandparents again. I won't be able to see my son grow up.

All I heard was crying and weeping in the background, my family holding on to one another in disbelief as the prosecutors and arresting officers stood up and left the courtroom.

I didn't know where I was going or what lay ahead. I couldn't imagine what my future would be like. I tried not to imagine it at all.

SEVEN

TERMINAL ISLAND

A week and a half later, two BOP guards came to my cell at 4:30 in the morning.

"Henderson, roll 'em up," he said. "Get all of your personal belongings and put them in this." He handed me a clear plastic bag.

I woke Carlos up and told him I was being shipped out. We'd traded addresses the night before. I didn't have a chance to say good-bye to my homies, but I knew I'd cross paths with some of them at some point during my bit.

Three guards took me and four other inmates down to a strip tank off the garage. They lined all five of us buck naked against the wall and gave each of us a serious cavity search. After we got dressed, they put us in shackles: chains that attached our wrists to our waists in front of us, and restraints on our feet. They loaded us in a van with two marshals, and another one tailed us in a follow car as we rolled out of the Detention Center and onto the streets of downtown San Diego.

It was the first time I had seen the outside world in eleven months. As we made our way along the 805 Freeway, all of us savored every moment, staring at the ocean, looking at all the different vehicles and the people inside them, eyeballing the women driving by. No one spoke.

When we arrived in Long Beach, I saw a sign that said Terminal Island and felt a little relief. Until that moment, I had no idea where the government was sending me. They had the choice of shipping me to a federal penitentiary in Arizona, Colorado, Oregon, or Terminal Island in San Pedro, just a two-hour drive from San Diego. Being locked up that close to home meant I could have visitors every week.

I saw the prison for the first time, a sprawling complex of two-story concrete buildings on a perfect rectangle of land in San Pedro Harbor, with only one road leading in and out. I saw the gun towers and a smokestack. My stomach started to knot up as we passed the Coast Guard station that was adjacent to the prison. I knew I had to pump myself up and be strong. I had to be a soldier. I thought about all the things Carlos and James had schooled me on for eleven months.

As we pulled up, several guards approached the van. "Everybody roll 'em up!" they shouted.

We inched our way out like penguins because of the shackles. I had gotten thinner in the last few months, so I was trying to poke my little bird chest out and flex my arms. We moved through the front entrance like a chain gang, flanked by four guards. I heard a lot of radio chatter from the guards' walkie-talkies as we went inside, and then the deep, echoing slam of the steel doors closing behind us.

We crossed the North Yard on our way to the receiving area. All the inmates had gathered to watch our arrival. All I saw was an ocean of faces—black, brown, and white—but it was the brothers I really paid close attention to.

As we turned the corner to go into receiving, I heard, "Hard Head!" "Yo, Big Jeff!" "Hendo!"

Niggas I hadn't seen in years were all calling to me from the crowd.

"Homie," one of them called, "we'll be waiting on you when you get out."

In receiving, they took my mug shot and prints, and a big Samoan lieutenant with a mole on his face asked if I was gang affiliated and if I had any enemies.

"No," I told him, "not that I know of." He had me sign a waiver stating that I didn't want to PC up.

My first cell was in J Block, where new inmates waited to be assigned their permanent living quarters. J Block was the oldest part of the prison, dating back to when the Terminal Island facility was first constructed in the 1920s. It was just like in the movies: a vast steel cage on three tiers. My cell was up on the third tier.

Everyone was segregated by race—blacks with blacks, whites with whites, Latinos with Latinos—and all of them just sat around staring at the new arrivals. There must have been two hundred convicts in there.

I kept thinking of what Carlos told me before I shipped out: "Keep your mouth shut, mind your own business, and do your own time."

My cellie was this dude Mike from L.A. A cool brother, he kept his cell as immaculate as Carlos had kept his and I knew I'd have to respect his rules since it was his house. After we made our introductions, Mike asked how much time I had.

I said, "I got a fresh nineteen and a half. What about you?"

"I got a dime to do. Anything you need, just let me know. This how we roll in this house here," Mike said. "Anytime you need to take a shit, let me know and I'll step out. Drop one, flush one."

When Mike got done telling me the house rules, I made my bed up, put my shoes under the bunk bed, and unpacked. I noticed he had a lot of pictures of his girl posted up on the wall and

I quickly looked away from them. Carlos told me not to fiend on pictures of another guy's girl. He also warned me never to put up pictures of my own girlfriends.

"Niggas will be lusting on your woman," he said. "Next thing you know, they'll steal your letters and start writing to them."

I put up some *Jet* centerfolds and kept all my private stuff in my locker.

As I was putting away my things, a couple dudes came to the cell. One of them was covered in tattoos and had a blue rag sticking out the back of his pants. I knew they were some Crip brothers, so I stood up and got ready to defend myself, because I figured they'd seen all my old homies from the Bloods yelling to me in receiving.

"My name's Lucky," the brother with the tattoos said. "People call me Lucky Luciano. Me and you have a mutual homeboy."

"Is that right?"

"Yeah, his name's T-Row. He said you were his little brother and that I should look out for you."

I said, "Yeah, man, that's my big brother. My name's Jeff but they call me Hard Head."

"T and me go way back. We were in Folsom together. He says this is the first time you've done a bit."

"That's right. I've been lucky."

Then he told me, "I seen them slobs greeting you when you came on the yard." He meant the Bloods.

"They were my folks out on the street," I explained. "They cool people, but I don't get caught up in gangbanging. I'm a hustler."

"Right, right. T's a good nigga. I got your back. You got any problems, you let me know."

After Lucky had gone, Mike was like, "Lucky's one of the top shot callers on the yard. He's connected. You ain't having no problems with that motherfucker on your team."

"That right? He was swoll' as a motherfucker." He was real buff; he obviously spent a whole lot of his time lifting weights.

Out on the yard, a lot of the homies were waiting on me. There was a guy, named Randy, I knew from back when I got stabbed in Long Beach. He was from the Rolling 20s, a Crip gang out of East Long Beach. Young Dave from the East Diego mob was there and so was Smokey from the Syndo Mob. There was a dude called Jamaica Man from Skyline who everyone thought was a Blood because he always wore a red cap, but Jamaica Man was just a Rasta.

"What up, Jeff?" Jamaica Man said. "We heard you were on your way out here. I got a care package waiting on you, snacks and necessaries. Something to hold you until you can get some money on the books."

I said, "Bet! Thanks, man," and he gave me a tour of the yard with the San Diego crew. On the streets of Diego, a lot of those brothers were from different gangs and didn't get along. But in prison that didn't matter because you grouped up according to the city you were from. Inside, Crip or Blood, if you were from Diego you were with the 619 card, which was named after the area code where we lived.

Then you had the 213, which was L.A., the 415 for the Bay Area, the East Coast brothers, and those from down South.

At nine, they started shutting the yard down, so I said "later" to the homies and went back to my house. I wanted to phone my family and tell them I was okay, that I was right over in San Pedro and that I'd get them all on my visitor list as soon as I could. I really didn't want to call my mother because I wasn't hard enough yet, so I called Dana and told her to pass the word along that I was doing fine, and that I knew a lot of people in with me.

After a month or so went by I was right in the prison mix. My first prison code infraction occurred with some Bay Area inmates.

We were in the chow line at lunch and I reached over one of the brothers.

He said, "Youngster, I wanna put you up on this. Don't ever reach over another man's tray. That's disrespectful."

I said, "I'm sorry, I was just trying to get something to eat right quick."

He told me again, "Don't ever reach over another man's tray. The dining hall is dangerous."

The dining hall was so tightly segregated, even the Jews had their own little pack over by the windows. There were guards for the main entrée, but you could get seconds on starch and vegetables. Everyone serving the food was very intimidating. They looked like linebackers doing life. Kitchen jobs were coveted because you could eat as much as you wanted.

I knew at any moment, anything could jump off in the dining hall, so I spent as little time as possible there. I hurried off the line with my tray and went to sit with the Diego crew. Out the corner of my eye, I saw a bunch of Crips coming in, mean-mugging and strolling hard. I tried to avoid eye contact but it was difficult, as they joined the black section, just three tables away. I was getting a little uncomfortable sitting there with all these Bloods.

"I'm going to get me some more vegetables and potatoes," Jamaica Man told us—he was on the healthy tip.

On the way back, he had to pass the Crip table. They made it all but impossible for him to get by without getting in their way. Jamaica tried to squeeze by, but he was so buff that he couldn't help bumping one of the Crips, this nigga called Al Capone.

Immediately, Capone jammed Jamaica up, saying, "Nigga, you just bumped me."

"I'm sorry," Jamaica said. "Ain't nothing personal. I was just trying to get by."

As I finish up, I'm nervously watching them with my one good eye and the tension was building. Capone sent this young brother from the Bay Area to our table to have words with Jamaica Man.

"Yo, Jamaica Man," he said. "Capone's trying to see you."

"See me for what?"

"He don't like how you disrespected him. This is the Crip's backyard, man. Why you wearing that red hat?"

"I'll wear any motherfucking hat I want, little nigga. Capone trying to see me? I'll see him. When and where?"

"South Yard."

After lunch, I got in Jamaica's ear, telling him not to go. "They'll jump you, man," I told him. "They got weapons."

On the outside, I'd always tried to stay clear of the Bloods/Crips bullshit. But inside, most of my homies were Bloods, and that made me a Blood in the eyes of most Crips. It was now a lot harder for me to not be on one side or the other. I still wasn't a Blood—and never would be—but I had to roll with my homies representing 619. One thing was for sure, if I ever had beef with anyone, I'd need them to back me.

Back in the unit, this OG dude named Love Bug let us know he had some heat for us if we needed it, but I didn't want to be caught holding a shank. Love Bug was huge, could bench four hundred pounds easy, and was the governor on the yard—and he was a gump. He would wear a T-shirt tied up under his chest like a woman and skintight dolphin shorts with his cheeks hanging out the bottom, and he always wore extra Vaseline on his lips. He was running his own store out of three lockers. If you were broke, he'd give you a pack of smokes, and you would have to give him back two. If you didn't pay him back, he'd make you let him give you head.

Jamaica Man was going to go see Capone by himself on the South Yard because he didn't want to bring us into it. When I saw him crossing the yard, I grabbed Smokey and some other homies from Diego and headed after him. By the time we got inside the weight room, Jamaica Man and Capone had squared off.

Then Capone started acting like he wanted to talk. When Jamaica put his hands down, Capone tried to sucker punch him but missed. Jamaica slammed a haymaker into Capone's chest and knocked him down.

Capone yelled, "Get him, cuz! Get him, cuz!"

A pack of Crips swarmed into the room. I took a deep breath and looked around for Smokey, but he was nowhere in sight.

This fool Big Easy, Capone's lieutenant, rolled up on me. He pulled his Coke bottle glasses off and tried to knock me in the head. He only grazed me, but I still ended up with a headache for three days. We were completely outnumbered, a bunch of Crips were stomping on Jamaica, so I yelled, "Po-po! Po-po!" as if the guards were coming in, and everyone scrambled.

I got Jamaica up and took him back to our unit. He was scuffed up, but nothing major.

Smokey comes up saying, "Where they at?"

"Nigga, where were *you*?"

"I told you I was coming."

"Man, it's done and over," I said. At that point, I lost all respect for Smokey.

Nobody liked Capone. He was an arrogant motherfucker. When word got around the next day, Jamaica was a hero. He didn't fuck with anyone, just ate, slept, and lifted weights. Even the Crips started to respect him. Hell, even some of the guards were pumping him up. A peace treaty was made. After that, Capone was finished.

A couple days later, after the tension eased, we were back on the weight pile, driving iron. I was the weakest of the crew. The most I ever pressed was one rep of 365 pounds. The rest of the crew pumped that like it was nothing. Crips and Bloods worked out at the same time, but at different stations. White boys, Asians, and Latinos all had their own times. It was very macho and competitive. I had to gain strength to defend against these big motherfuckers in case anything went down again.

The 3:30 horn rang to let everyone know they had thirty minutes until the next count. As everyone made their way back to the unit, I still had three or four sets to do, so I stayed behind. This fat black guard just out of the navy came over trying to prove himself.

He starts yelling at me, "Count time!" but I kept driving to get my last set in. He stands over me and says, "Inmate, put the weights down. Turn around."

"Turn around for what?"

"For disobeying a direct order."

"I still have fifteen minutes left."

"Turn around. You're going to the hole."

I yelled to one of the homies to get my bedroll and personal items and hold them for me so that no one would steal them while I was gone.

We went through a series of sliding doors. It was very dark. There were several senior officers and regular COs.

"Henderson," one of them said. "What you do, man?"

"I didn't do nothing. I was getting my last set on the weight pile and your guy over here swears I was refusing a direct order."

"All right, Henderson, welcome to SHU." He asked another guard what they had open.

"Put him in there with Lucky. I'm sure they have a lot of war stories to trade."

After the strip search, they put me in a dark brown jumpsuit with SHU on the back. All the Special Housing Unit security procedures were new to me. They had two officers on me. We went upstairs, down a long hall, and the guard knocked on the door.

"Lucky, I got you some company."

"Who's that?" he said, jumping up from his bunk.

"Inmate Henderson."

When he saw my face, he said, "Aw, bet. T's little brother."

Our neighbor down the hall was Jeffrey McDonald, the former Green Beret captain serving life for the 1970 slaying of his wife and two children. He was well groomed and polite, didn't seem like a killer to me. He always got his attorney visits in a private room and everyone said the lady who visited wasn't really his lawyer. She wore short skirts and word was he was fucking her in that private visiting room.

I spent a week in the hole before I was brought before the DHO, Mr. Page, a short stocky black man. I'd heard nothing but good things about him, that he wouldn't railroad anyone, and yet he found me guilty of refusing a direct order. I got time served and an infraction on my record.

When I got out of the hole, my caseworker, Mr. Welsh, hooked me up with a job mopping and sweeping the TV rooms three times a day. It was cool because I could watch TV and had lots of extra time to write letters and kick it with the homies on the weight pile—talking about all the money and cars we'd had on the streets.

I wasn't even thinking about trying to get my life together. I didn't give a fuck. The white man had just given me nineteen and a half years. I wasn't trying to work with the Feds, wasn't trying to get a better job. I just wanted to lift weights and kick it with my homies.

So I did my little job, lifted weights, and watched TV. We watched *Rap City* and we could never wait to watch *Midnight Love* after the ten o'clock count. But no matter what, we always had to watch whatever the shot callers in the TV room wanted to watch. The soap operas, the *stories,* were mandatory, every fucking day:

One Life to Live, All My Children. There were guys in there who'd been watching those shows play out for the last twenty years.

I never once touched the television, which was another one of Carlos's rules. "Never touch the TV, and always sit with your back to the wall in case a squabble breaks off and motherfuckers start slinging chairs."

I'd been mopping the TV rooms for six or eight months when Mr. Welsh called me in and told me a job on the Cadillac crew was opening up. The Cadillac crew was an easy job, fifteen or twenty minutes a night sweeping up cigarette butts in a little section of the North Yard. The job was sweet. I told him it was perfect for me because I was planning to start attending school in the daytime. That was bullshit. I just didn't want to do any hard work. I wanted to be lazy and kick back.

I got on the Cadillac crew and started living large, as good as any inmate could. I had about $5,000 on the books, which lasted me for a while because we were limited to spending $125 a week. Dana came to see me every weekend. Moms only came about once a month, but that was cool because it always made me too emotional to see her. Dad came one to two times a week because he lived just thirty minutes away. Dad always showed up an hour early in order to be first in line at the vending machines in the visiting room. Inmates couldn't get anywhere near the machines, but their families would empty them out almost as soon as visiting hours started.

My favorite was the tuna fish sandwich on wheat bread with apricot juice and barbecue mesquite chips. I'd settle for a bean burrito or a pita pocket when they ran out of tuna sandwiches. For me, eating food from those machines was like going to a restaurant.

Dad would spend an hour or two briefing me about everyone in the family and the outside world, and I would sit there lusting after the female visitors. I was tired of looking at men 24/7. The same fifteen hundred faces 24/7. The smell of funky asses and feet day in and day out.

The only thing I didn't like about having visitors was being strip-searched before and after. I could see them stripping us coming back, but why did they search us on the way into the visiting room?

One of the guards who worked visiting room duty was Antwone Fisher. Officer Fish, as we called him, had come from the streets like most of the brothers in the prison, but he'd gone into the navy and then joined the Feds. Years later, he turned his life story into a book and then a movie that Denzel Washington eventually starred in and directed. Officer Fish was a cool brother who understood the life we inmates came from. All he wanted to do was his job. He never put any pressure on anybody and he never made trouble. He looked the other way if you were touching up your woman. He even skipped the strip searches if there were only two or three guys in the group. Hell, I am sure even the guards got tired of looking at assholes.

I asked Officer Fish why they searched us on the way into the visiting room, and he told me that an inmate had once smuggled a weapon inside and beat his woman down.

On Friday nights, Dana would get a motel room near the prison or stay with my grandma so she could spend all of the weekend visiting hours with me. I always had a fresh haircut and a fresh shave and was lotioned up with Smell Good. I'd trade the guys in the laundry cigarettes for new khakis whenever she was coming. Dana always had that baby-doll dress on. She would get there three hours early to get the best of everything. Sometimes I wouldn't eat for a day after, I was so full from vending machine food.

Dana and I never went too far in the visiting room because I'd heard rumors about guys losing visiting rights for a year for trying to have sex. So I learned how to take care of myself. I also never put any other girls on my visiting list because Dana spent so much time in the visiting room that she knew everyone in it—all the guards, the families, and everybody's woman—and so she would find out. Lucky once had a girl visiting him when his wife showed up. The women started fighting and Lucky lost his visits for a year.

In prison, everybody had an angle, a way of making the time pass. Everyone did their time their own way. Some guys were into drugs, some guys were into guys, some guys gambled, played basketball, reminisced about the streets. Some guys did their time studying history. Some found God and used religion as a crutch. All of them wanted to recruit me: Black Muslims, Christians, Five Percenters. They all wanted to make their cards bigger.

I wasn't trying to get with any of them. The 619 was enough for me. Everyone had their own tailor-made blueprint for doing time, and my blueprint was to stay out of everyone else's mix. I'd get up at 5:30, eat, hit the weights, take a nap before lunch, and walk the track after lunch for two or three hours with one of the homies down on the South Yard. I'd do my little job with the Cadillac crew in the early evening.

In the evening, I wrote my love letters and caught *Midnight Love* on BET. All the brothers would sit there like zombies, watching the videos to those romantic slow jams. After the videos, all the homies would run to call their girls and try to get some phone sex in. I stayed clear of that. Beside being nasty, it was a tense scene and it was dangerous to be near the phones after *Midnight Love*.

Another thing Carlos taught me was, "Never call a woman early in the morning or late at night and you won't ever get your heart broken. Always give her enough time to get home so she could tell you she stayed in all night."

When the brothers would hit the phones, I would hit the shower with Dana's photo, or sometimes even Carmen's. It all depended on who I missed emotionally at that moment. In my mind, I was out of prison and making love to each of them. When I'd get back to my bunk, I'd stare out the window and watch the moon and stars over the Pacific, and tune my Walkman to KJLH so I could listen to love songs until I fell asleep.

I'd been at Terminal Island about a year when the prison chaplain, Father Bill, called me into his office one morning. He was a white man with a full white beard. Dana had been pressuring him to approve our marriage. Her ultimatum to me was to marry her or she'd start looking for another man. I loved Dana, but I didn't want to get married. Still, she'd been a soldier for me through my arrest and trial and right up to that moment. And she had been there for my mother.

A lot of the homies were saying, "Don't get married, man. It's a jinx."

But I figured, what have I got to lose? Hell, even if she only sticks with me through half of my sentence, that's still ten years' worth of visits.

The wedding ceremony was held in the visiting room. My father, sister, and my auntie Barbara were all there. Moms wasn't. She wasn't too supportive of my getting married. All she said was, "At this point, you have to do what you have to do." There I was in federal prison serving twenty years, and she still thought of me as her baby boy.

A couple weeks after we were married, Dana started taking extra days off to visit me more often. One particular Friday night, we had a good spot with a view of the ocean, but a few Mexican couples next to us were getting out of control. Me and Dana were like, "Damn, maybe the cameras aren't working tonight."

At 9:30, an officer gave the fifteen-minute warning and everyone went back to their units. After the 10:00 P.M. count, one Mexican boy who'd been fooling around with his girl in the visiting room started going into convulsions, but none of his homeboys were calling the guard.

I was, like, "Somebody call the guard! That motherfucker's dying!"

One of my homies took me aside and explained that the kid had a balloon of heroin in his intestines and his friends were waiting for him to shit it out. It turned out that his girlfriend had passed it to him mouth to mouth while they were kissing, but she'd put the heroin in a party balloon instead of a surgical glove and it had sprung a leak in his stomach. His friends were more worried about getting high than saving his life.

I shouldn't have been surprised. It seemed like every time I went to the bathroom at night, if guys weren't fucking, they were shooting heroin with guitar strings into any vein they could get at: between the toes, in the neck, even right into their dicks. You'd think they would have the courtesy to at least clean up the blood they'd left behind, but they never did.

The guards showed up maybe ten minutes later, but it was already too late for the Mexican dude. He died in the infirmary that night. In all my years of dealing, I never fully realized until then just how far people would go to get high. Drugs were just business to me, a way to get the things I wanted. I just wanted to feed my family better. I wasn't about the gangster shit, I wasn't killing anyone. Anyway, that's what I'd told myself. All that time I'd been blind.

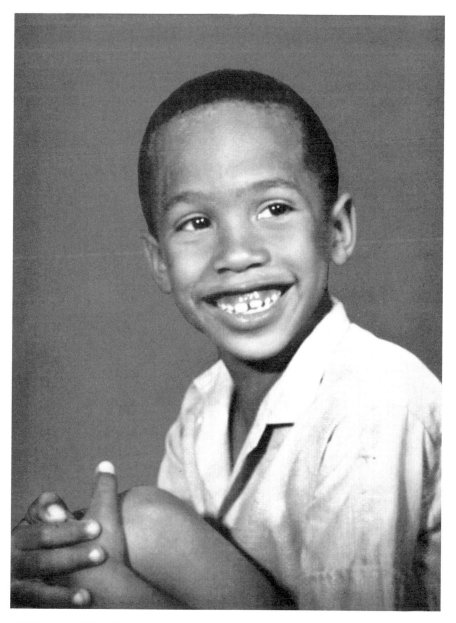

"Little Hard Head"—I was a young thief
in disguise behind this big smile.

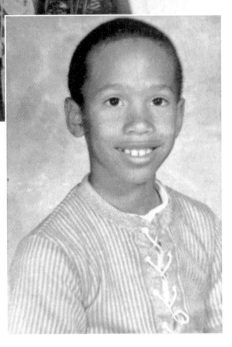

**Finding ways to stay in
and out of trouble—**
Eight years old *(right)*;
graduating from high school
(above), already a young thug.

My first taste of real money—Twenty years old,
dealing crack, and playing out at Caesars in Vegas, 1983.

The spoils—*(clockwise from top)*
'86 custom Nissan truck, first prize in
a supercustom car show in San Diego;
'68 Cadillac Coupe de Ville, my first
car, bought with $375 in weed money;
my pearl white '85 Mercedes 500 SEC,
fully dressed; '78 ragtop Coupe de Ville.

Before the fall—Riding in style to my twenty-fourth birthday party in the grand ballroom at the Omni in San Diego. I knew it would be my last birthday before prison; I dropped $10,000 on the party. I give the presentation high marks.

The big house—Terminal Island, San Pedro, California. In the top-right corner is my mug shot, taken when I first arrived.

My first years inside—
Hitting the weights helped pass
the time; I tried to stay out of
the mix the rest of the day.

A sight for sore eyes— with my parents during visitation.

My calling—After several years in prison kitchens, I had found a path to redemption.

Into the unknown—As I walked to freedom on October 2, 1996, I was nervous because my redemption had yet to be tested on the outside. I was lucky to have a strong support network: *(above)* my son's uncle Larry, my sister Charlean and her daughter (with rose), my son Jamar, and my future wife Stacy; *(left)* with Stacy.

Chef Jeff—In my first position, posing with the other chefs at the Marriott Coronado *(above)*. The vision of my prison dreams had been manifested—I was running my own restaurant *(right)*.

In the kitchen—During my first month at Caesars Palace in 2000, I ran into conflict when I wouldn't condone some unethical activities taking place. The crude drawing *(above right)* was part of an attempt to run me out of the kitchen. Instead, I was promoted to chef de cuisine—the first African American to hold the position *(top)*. In the kitchen at the Ritz Carlton, Marina Del Rey, with Paul Bocuse—the legendary French chef *(above left)*.

"It's all about choices"—

On the South Side of Chicago with high school students, sharing my "Twelve Powerful Steps to Chefdom." It's special to return to the inner city to inspire kids who come from my background. My message: stay passionate about your dreams.

My inspiration—

With the family *(right)*.

Tough love—Preaching about the dangers of the
streets *(top)*. Fifty-six grams of synthetic crack *(bottom);*
on the street it would be worth $3,000 and ten years
to life in the pen. I tell them that it's not worth it.

The executive chef—Finally reaching the
pinnacle as the executive chef of Café Bellagio.

EIGHT

~~~~~

HARD HEAD NO MORE

Even though I'd been ignoring and avoiding all the different prison groups, some of the things the Black Muslims talked about caught my interest. They said that the black man had built civilizations, kingdoms, before we came to the wilderness of North America. A part of me thought *I'm not with that shit.* But I really started to think about it when I was in my bunk at night and I'd talk about it with Dana during our phone calls and visits.

Before long, I attended regular meetings with the Black Muslims. My job on the Cadillac crew wasn't very demanding, so I had a lot of leisure time to listen to their teachings. And I found myself wanting to share what they were preaching with Dana. But her people told her, "Don't listen to none of that jailhouse rhetoric. He's fucking your mind up."

Her friends and family were already starting to hate on me. This was just one more reason, and I knew she was listening to them. We'd only been married six months, but I was already getting suspicious about what she was up to on the outside. She started getting in shape and doing her hair up differently. She stopped wearing the little baby-doll dress to visits, so I couldn't touch her up that much. While the signs said I was losing her, I didn't quite let myself believe it. It would hurt too much to believe it. Besides, more and more of my focus was the Nation of Islam lectures.

I knew in my heart I was Christian, and I loved the Lord, but I was angry at him. Jesus allowed these white folks to give me nearly twenty years, to take me away from my family. I didn't know then that he had a bigger plan for me. My mind was still undeveloped, and my soul was still weak. Moms always said God works in mysterious ways. I guess I had to go through this experience to come out the other side.

There was one particular Black Muslim who I really began to listen to. A reformed Crip from L.A., Eddie X was a jet black brother who looked like Idi Amin. Every time he saw me, he'd say, "Brother Jeffrey, how you doing?"

I'd say, "What's up, Eddie X?"

"Slow motion, man. Ain't nothing happening" and he'd add, "Hey, brother, I want you to come over to the mosque on Sunday."

"Yeah, man, I'll be there," I would tell him. I didn't want to watch a lot of tapes and listen to these brothers talk about hating the white man. They put everybody down, blamed everyone for the black man's problems.

Another brother by the name of Kevin X caught my attention. Kevin X was a longtime soldier in the Nation of Islam. He was part of what they called the First Fruit, people who had joined the Nation while founder Elijah Mohammed was still living. He'd been a personal driver for Louis Farrakhan and a right-hand lieutenant under Kahlil Mohammed. Kevin X was doing fifteen years for armed bank robbery. He wasn't a large man, but there was an aura about him that you feared. There was a sense of danger around him.

He was at constant odds with the prison's Orthodox Sunni sect, which was headed by a six-foot-five, 240-pound imam named Stuts who was serving life for strangling a man to death in Guam while he was serving in the navy. The Sunnis embraced Malcolm X but the Nation rejected him for rejecting the teachings of Elijah Mohammed after his pilgrimage to Mecca.

One Sunday after a visit with my father I decided to see what those Nation brothers were really talking about. I wasn't going to commit to anything, but they had me curious, and I needed something to keep my mind off the fact that Dana wasn't visiting me too often anymore.

Eddie X held class for two hours every Sunday in the library. He was the acting minister and a couple young brothers named Mustafa and Johnny 6 X assisted him. There were about twenty-five brothers in there, listening to the teachings of Louis Farrakhan through Eddie X. After Eddie finished preaching, they popped in a videotape of Farrakhan.

The first tape I saw was an old one of Farrakhan in his younger days, when he used to preach the fiery rhetoric that the white man is the devil with blue eyes and blond hair. I didn't understand what he was talking about, but as he started preaching from a historical standpoint, I began to listen more and accept more. I started to feel that my fate at the hands of an all-white jury was just the tip of the iceberg.

A white jury gave me nineteen and a half; the white man brought drugs into the country; the white man enslaved us; the white man systematically mixed his blood with the black man. At the end of the tape, I was moved. My thinking began to broaden.

I started having self-talk, asking myself: Why? Why after four hundred years in this country are the black man and woman still oppressed, depressed, broke, imprisoned, and held back?

After the tape, everyone stood for prayer. I never closed my eyes because, as always, I wanted to observe the others. They held their hands palms up like they were opening the Koran and said an Islam prayer:

In the name of Allah the most merciful, I bear witness that there is no God but Allah.

It made me think back to when I was fighting my trial, when I kept praying to Jesus and to God. In church they preached that God is Jesus and Jesus is God. Now Islam was preaching that Allah is God. I couldn't understand why there were all these different names for God, and why we all prayed differently. I was confused.

All my life I was only exposed to Christianity. I didn't know anything about Muslims, Five Percenters, Black Nationalists, Jews, atheists, Jehovah's Witnesses. Of course, I knew there were other religious groups, but I knew nothing about them.

As the service broke up, Kevin X said, "I'm glad you made it, Brother Jeffrey."

"Yes, sir," I said. "It was very interesting."

"I want you to take this."

He handed me a book. I read the cover: *Message to the Black Man* by the Most Honorable Elijah Mohammed.

At that point in my life, I had never read a book cover to cover except for *Curious George* and *James and the Giant Peach*.

I put the book in my locker, went and took my shower, and then hit the sack. But I couldn't sleep that night. My mind was weary and my heart heavy because things were getting colder and colder with Dana. I needed a crutch, something to help me do my time. I needed to clear my mind and strengthen my heart.

Flicking on my night-light, I opened the book and began to read.

I didn't understand a lot of the jargon, but one thing that struck me was how Elijah Mohammed kept repeating "Black

Man." The Black Man was powerful, intelligent. The Black Man was on Earth before any other mankind, had created the first civilizations. I'd never before seen books that depicted black heroes and black sheroes.

I was blown away, mesmerized to know that we were great people. I remembered that, as a child, I didn't understand why my hair was so nappy, why my lips were so big, why I looked so different from all the white kids on TV.

The more I read, the more questions popped into my mind: Why did blacks get their hair straightened? Why were African names taboo in America? Why was I called Jeffrey Henderson? What was my true name? The name *my* forefathers had before they came to America?

I fell asleep with the book next to me. In the morning, the battery on my night-light was dead.

The next day I was taking a nap on my bunk when Kevin X sat in the folding chair next to my bed and tapped me on the shoulder.

"Brother Jeffrey," he said, "I want to read something to you." He'd brought another book called *How to Eat to Live* by Elijah Mohammed. He'd seen me in the chow hall feasting on pork. I loved that swine. One of my fondest memories of life on the outside was when Carmen used to cook us those thick center-cut pork chops with macaroni and cheese and sweet peas. I listened to Kevin X, but I wasn't about to give up pork.

Kevin X became a regular in my unit, giving me private lessons on the Black Muslim philosophy. My homies from the 619 started noticing the change. I stopped hanging out with them except to lift weights. I stopped walking the track with them, stayed out of the TV room. I was hanging with the Black Muslims—discussing current events, debating the fate of the black man in white America, and just listening.

I wasn't sold on all of their ideas, but they became my friends and my brotherhood. They were always respectful and polite. They called me "Brother," and we always answered one another, "Yes, sir."

They were the most respected brothers on the yard. No one messed with the Nation, not the gangbangers and not even the guards, because members of the Nation stuck to themselves and never brought any drama.

The only person who had a beef with the Nation was Al Capone. He didn't like Kevin X because Kevin had converted one of his Crips. The mission of the Nation was to convert Crips and Bloods into soldiers for the Black Muslim cause. Kevin was a master at it. He would tell them: "How can you kill someone who looks like you, talks like you, and walks like you? How can you claim a neighborhood that you don't own? How can you claim a block that's not yours? How can you build an army to go against your oppressed brother? How are you willing to die for a color, a flag that represents ignorance and self-pity? How can you do the work of the Ku Klux Klan?"

That kind of talk would change some minds, but not most of them. A lot of people didn't want to hear any of that. It was just noise to them. A lot of the homies I had been hanging with started sweating me about my affiliation with the Nation, so I took it as an opportunity to share my newfound knowledge of self.

They were like, "We don't wanna hear that. Them niggas is crazy. They be out in the sun, one hundred degrees with they shirts buttoned up. You been listening to too much of that bullshit up there in the mosque."

I defended the Nation because I'd begun to feel myself developing intellectually. Reading had always been a struggle for me, but now I was reading complicated passages relatively easily and drew pleasure and mental strength from it. I felt stronger, wiser. Throughout my life, I had difficulty expressing complicated thoughts, but the Nation allowed me to do so on a regular basis and I was becoming more and more confident in my ability to have complex conversations. Even my family started recognizing these changes in me.

Soon, though, the Muslim brothers started pressuring me to write a letter to the Nation of Islam headquarters in Chicago to ask for my X. That would make me an official member—a member for life. While I totally embraced the social component of the Nation's teachings—love yourself; be strong; respect your woman; value education, your brethren, your family—I had problems embracing the religious aspect of the Nation's teachings. I was still the same man my mother had raised. I was still a Christian at heart and could never abandon that.

There was another community on the yard that was headed up by a wise old man named Imze. He'd been a member of the Black Liberation Army and he would get on the yard and counterteach the young brothers who the Nation was trying to convert. Imze's social message was similar to Elijah Mohammed's, but his religious beliefs were totally different.

"We believe in the spirit," he would tell his brothers. "We worship the sun, Ra. We worship life; we worship the water. We worship nature."

Imze taught me a lot more about the history of our culture than the Nation did. He introduced me to a book called *The Destruction of Black Civilization* by Chancellor Williams. It talked about how the Arabs had invaded North Africa and forced Islam on the Africans just like the Europeans had forced Christianity on us.

Another book he gave me was really deep. *They Came Before Columbus* by Dr. Ivan Van Sertima argued that Africans had come to the Americas on boats way before Columbus and Vespucci, and how they had traded with the native people.

While I was wavering between the Nation and Imze's sect, I became involved with prison organizations. I helped facilitate programs for Kwanzaa and Black History Month. I joined Toastmasters International to improve my speaking skills, began reading in the law library, went back to school, and earned my GED. I cofounded the Black Cultural Workshop, a think tank where on Sundays we invited the best minds in the compound from all religious and racial backgrounds to discuss philosophy, religion, and current events.

I even volunteered for the inmate suicide watch. When inmates became unstable, they were put naked in the suicide tank. They reminded me of the guys going crazy in the hole. I felt a lot of compassion for them because I remembered the howling all through the night during my time in the hole. Some guys were doing life. Some guys had been in PC for years. I identified with them. They made me remember when I first fell, all the shit that went through my mind: death, escaping, and depression. I'd sit by the door and talk to them, telling them it would be all right.

"You have family that cares about you," I always told them. "You have people who love you. It's not the end of the world."

Dana and I had been married for a year. She was skipping more and more of our weekend visits and our phone conversations were shorter and shorter. Half the time, she wouldn't answer the phone at all. She was getting tired of the run.

In reality, she was doing my time with me. She had put a hundred thousand miles on that Hyundai of hers. She had $300 phone bills every month from my collect calls. She was a young woman and I was breaking her down. Her family had no love for me. Her father and mother saw that their daughter was in love with a criminal doing nineteen and a half years.

I confronted her on her next visit.

"Dana, what's up?" I said. "It seems like you been tripping lately. You been messing with somebody out there?"

"No, I ain't messing with nobody! I'm just tired, Jeffrey. You're stressing me out. I never have money anymore, my car is running raggedy, and I'm taking pressure from my family and all my friends."

"You're the one who wanted to get married," I said. "I told you to leave me when I first got here, didn't I? I told you to leave me, Dana. I told you to set me free. I've got twenty years to do. I told you it was gonna be tough."

"I know," she said, "but I loved you. I wanted to be there for you. I wanted to be down for you. But I don't think I can keep going the way I'm going."

"Then why did you commit to this? I could do this time by my motherfucking self."

"Don't talk to me like that, Jeffrey."

I knew she was hurting, but I also knew that she had already moved on. Word had gotten back to me that her friends had hooked her up with some straitlaced dude. I hadn't planned to mention it, but I felt so hurt and alone then that I wanted to strike out.

"I know you hooked up with that motherfucker," I said. "But you know what? That's okay, because I know I'm gonna be home one day. I'm gonna make it. I'm gonna get mine."

We just stared out at the ocean for the rest of that visit. It was over and we both knew it.

She quit me cold turkey after that, just disappeared and changed her phone number. It was the best way. Carlos had warned me: "Do the time by yourself, youngster."

After eighteen months of being down, my number came up to be transferred to the South Yard. It was a plush inmate housing community nestled under palm trees with green rolling lawns and the best ocean view in the prison. The floors were waxed and polished. It was reserved for inmates who showed signs of rehabilitation and were on good behavior. The Carnival Cruise Line ships would pass by just a hundred yards away. I was still in prison, but it was more civilized.

The inmates were different, too. There was no gangsta bullshit. Most of the tenants were older men with mob ties or Wall Street moguls. There weren't too many brothers, and the ones who were down there were mostly the black leadership.

My new cellie was a homeboy named Brady from San Diego. Brady was a cool dude; he had caught his case with two of his brothers. I was glad to share a cell with him, at least because he kept the cell immaculate. It was a lot different from all the other cell blocks that I had lived in. The sky-blue painted cell had wooden bunk beds, with decent-quality mattresses, a small bookshelf, a closet, and a nice window with bars, but there was a view.

Down the hall, next to the bathroom, lived this weird older white man. His name was Larry Layton, a small, tightly built guy with graying hair. He never spoke to anyone. In our unit, he would walk the halls at a deliberate pace. Sometimes he'd just stand in the yard staring at the ocean in the green flight jacket that he wore no matter what the weather. He'd been the triggerman in the 1978 murder of Congressman Leo Ryan on a Guyana airstrip when Ryan had come to inspect The People's Temple at Jonestown.

Across from us was where the mob lived. There was Rosario Gambino, the NYC crime boss serving forty-five years for his part in the "Pizza Connection" heroin ring. Mr. Gambino was in his late forties or early fifties, always kept his hair slicked back, and wore a presidential Rolex and a thick gold necklace with a Virgin Mary medallion.

We became acquaintances right away. Mr. Gambino worked out on the weight pile twice a day, walked the track for an hour each night, took naps, and cleaned the TV room. He was a real Old World type of gentleman. When his wife and family visited, he spent thirty minutes in private conversation with his sons before the rest of his family came over.

Mr. Gambino spent a lot of time reading and writing screenplays in Italian. After a few months, he asked me to read him some stories from the *New York Times* because he didn't read English very well. He never talked about his case, but I'd always been interested in why mobsters of his type were so notorious. Maybe it's the way they operated covertly, while brothers did their shit openly and foolishly.

They weren't the only ones.

One night, I went to the bathroom at about 2:30 in the morning. I was walking kind of sleep drunk and in a daze. But suddenly I was alert when I looked over to one of the stalls and saw four feet in it. I heard some grunts and hurried back to my cell. I knew that type of thing was going on, but it was a shock to hear it—and almost see it—for the first time in my life.

"Man, that ain't nothing," Brady told me. "I went in one night and there was a dude in there douching."

"Is that right?"

"Another thing tripped me out," he said. "You know that old man who's always feeding the cats?"

"Yeah," I said. "I know him. He's a real nice old man."

"Well, that nice old man got busted by a couple inmates. He was taking the juice from a can of tuna fish and spilling it all over his Johnson and having the cats lick it up."

I went crazy.

"What you want from him?" Brady said. "He's doing life, man. Motherfucker'll never get a blow job again."

There had, in fact, been cats all over Terminal Island for the first year or so that I was there. They meant the world to a lot of the inmates. Not for nasty shit like that old man; it was because they could touch them and it was like touching freedom. Those cats could go into the free world any time they wanted, but they chose to hang out with us. When they started overbreeding, the warden had them removed, and a lot of the inmates were devastated.

After eight months or so in F Unit, I was hanging out in the library so much that I started ignoring my job on the Cadillac crew. Several days went by and I hadn't been tending to my little section on the North Yard. One night, I got called to the lieutenant's office and he fired me from the Cadillac crew. I pleaded with him to let me keep the job because it allowed me the freedom and time to read and study, but there were too many inmates who wanted the job and wouldn't fuck around if they got it.

The lieutenant told me, "Report to the kitchen at 6:00 A.M. for extra duty on the pot and pan crew."

I wouldn't dare challenge him, but just said "Fuck it" under my breath.

I'd heard a lot about what went down in the kitchen, and I didn't want to be tied down eight hours a day. I was, like, "Shit, the only time I ever washed dishes was when Cali Slim would cook dinner for Moms and me or when Moms put me on punishment."

I couldn't imagine washing pots and pans for fifteen hundred inmates three times a day. My idea of a kitchen was being at Grandma's waiting to eat the food that I loved. Now I was going to be locked up in this massive industrial kitchen full of steam and grime. It never entered my hard head that any good could ever come of it.

NINE

~~~~~

KITCHEN HUSTLE

The next morning, just after the 4:00 A.M. count, I was awakened when the overnight guard shined his flashlight in my eyes.

"Henderson," he said. "Kitchen duty."

"All right, sir."

I hadn't been sleeping soundly anyway—I never slept soundly. I'd wake at the drop of a dime. I knew better than to sleep hard in prison; you just never knew who might creep up on you while you were dreaming away. Slowly, I got dressed, brushed my teeth and washed my face, and made my way to the front door of the unit. Several guards were there to escort inmates to the kitchen for breakfast duty. There were several different guard positions at the prison: The captain, who was in charge of security, oversaw all lieutenants who in turn oversaw the correction officers, and then there were the unit guards, escort guards, perimeter guards, special housing guards, maintenance guards, and food service guards. This last group was made up mostly of guys who had at one point worked as cooks in the military, and they were the ones who marched us across the North Yard that morning. *Shit, I really fucked up this time,* I thought, and I knew that if I messed up again I'd wind up in the hole for a long while.

We were received in the kitchen by the food service guards. A tall skinny one named Parnell gave me a quick briefing about my job duties. *Whatever,* I thought, as he showed me the pot and pan room where I'd be working. There wasn't much action going on at first. I just sat around the dining room doing nothing with the twelve other inmates who relieved the overnight detail. Then the 6:00 A.M. horn blew and the breakfast rush rolled in.

"Henderson," Officer Parnell said, "let's get to our area." I jumped up and went over to the dish area.

Then it began. One of the Mexican boys showed me the system they had. I'd be working at a three-compartment sink. One man scrubbed the pots, the second inmate rinsed them, and a third would run them through a sanitizing agent. A fourth guy on the end took the cleaned and sanitized pots and pans and put them up on racks.

They started me on the rinse area. Everybody seemed to be pretty cool. There were three or four brothers in there who started rapping and singing as soon as the work started coming in, and I fell right into the mix, banging those pots out. The only thing on my mind was keeping in rhythm with the pot and pan line, to show the other inmates that I could flow with their system.

About an hour into the job, an inmate hit the back door. He had a whole pan of bananas.

"Damn! We get these?"

He said, "Yeah, these are some extra ones we just got from the guards."

"That's what *I'm* talking about," I said. "But what about some chicken? The bananas are cool, but I could use some of that yard bird, man."

The inmate said, "Slow down, youngster. You're new, you'll get yours in due time. Be happy about the damn bananas!"

"All right," I said, and ate one on the spot while I kept up with the rinsing, shoving a couple more in my pockets.

Doing the dishes wasn't all that bad, except there seemed to be an endless flow of them and my hands were getting banged up. The guy doing the scrubbing was older, and going really slowly. And the slower he went, the more the pots piled up. At the rate he was going, I'd be stuck in that hellhole all day. I wasn't up for working all day at a steady pace; I wanted to get done and get out of there.

So I said to the old man, "Why don't you get on rinse and let me take over scrub? I got a lot of energy." As soon as we switched, I got right with it, scrubbing those pots up boom-boom-boom and slinging them into the rinse water.

We were moving, because I was working hard. But everyone looked at me like I was some youngster who didn't know any better.

"I'm not trying to be here all day," I told them. "I'm trying to get my eat on and then hit the weight pile."

A little while later, without slowing down a bit, I asked about the bananas.

"All you gotta do is hold tight," the old man said. "The guards make sure we eat real well, as long as we make them look good in front of the warden. All we gotta do for that is take care of this dish room—make sure all the pots and pans are clean and that the room stays organized."

"That'll work," I said, thinking it was a fair trade-off.

But by the end of breakfast at 7:30 there was an ocean of pots and pans that still needed cleaning, and I was back to telling myself that there was no way I could keep on being a pot man. Then Officer Parnell came in and handed around cinnamon rolls and more bananas. I was sold. I still hated the scrubbing, but I was starting to catch on that the perks of being in the kitchen were worth more than just eating better.

"This is cool," I told the old man. "This is how you guys be selling all that food on the yard?"

"Yeah man, but keep that on the low," he said. "We always get the most leftovers at breakfast because most of the guys sleep in and just wait for lunch."

He explained that the guards calculated how much food to order for each meal based on the counts. If they counted fifteen hundred inmates, for instance, they ordered fifteen hundred bananas.

"So if three hundred guys don't show up for breakfast," he said, "that's three hundred extra bananas. Those go to the kitchen crew."

Bananas were a delicacy in prison. Everyone was always trying to get healthy and bulk up, so if you had some bananas and cereal and milk in your locker back at the unit, that was a great snack. Out on the yard, you could easily get $2.50 for a banana.

We wrapped on the pots and pans at midday. I thought I'd have to be there for eight hours a day, but I realized it was really only going to be six. A whole new crew came in for the lunch detail and I started to explore the rest of the kitchen. As long as I did my job and was at my bunk for all the daily head counts, I could spend the rest of the day pretty much however I wanted, and could check out unrestricted areas like most of the kitchen.

It was about half the length of a football field, just huge. I'd never seen anything like it in my life. I walked down a long corridor, looking into all these different kitchens and rooms.

My first stop was the receiving area, where all the food and supplies came in through the sally port, a fenced-off loading dock where everything was gathered for multiple inspections. No boxes, no cans, no containers, got into the kitchen without first being inspected for contraband. The inmates were never allowed near the supplies when they first came in from the outside, because if they knew where the food was coming from, they could arrange a smuggling operation with someone on the other end.

After receiving came the first big kitchen, the bakery. There weren't any brothers working the bakery; the white boys had it sewn up tight. They'd been there for a long time already, and no one ever left the bakery since it was one of the most coveted jobs in the kitchen. The bakery churned out doughnuts, maple bars, twisters, bear claws, and cakes and cookies, as well as some special items that never hit the chow line. My favorites were the cinnamon rolls. They were enormous and buttery, with icing and brown sugar, and were laced with maple syrup. The cinnamon rolls were special because, like doughnuts, we only got them on Sundays. For some reason, the Feds always fed us the best foods on weekends. A lot of times I'd even skip the visiting room vending machines because I'd filled up on so many pastries at our jailhouse Sunday brunch.

I really wanted to be down with these bakery guys and gain access to their extra sweets. So I walked into the room and said, "Hey fellas, what's up?"

They kind of nodded at me. I could tell they weren't fucking with no brothers. On top of that, I was a new face in the kitchen, so I just stood back and watched them work.

I was very curious about all the machines in there. One of them was the sheeter. They fed it round mounds of dough on a long electric belt. The machine would knead the mounds into flat, square sheets of dough, which the inmates would then fold and feed back through the machine to make the sheets even thinner. The bakers then spread the sheets out on a wooden table dusted with all-purpose flour, then cut doughnuts with a ring mold, and put them into a proofer to rise. Once the doughnuts had risen, they were put into the deep fryer. After a few minutes in the fryer, the bakers would flip the doughnuts with long wooden sticks to cook them on both sides. I'd never seen that before, and it fascinated me.

The huge oven looked like something you'd see in a crematorium. An inmate opened the heavy steel door to reveal six long shelves that held four sheet pans each, filled with cinnamon rolls, cakes, and cookies. I'd later learn that all of the baked items that were served in federal prisons were prepared according to military recipes. The traditional pies served were Boston cream, lemon meringue, apple, and peach, all made with canned fruits that reminded me of the government commodity food I'd eaten when I was growing up, but the pies were still very good.

On the other side of the bakery, a guy was running the giant ninety-quart Hobart mixers. I was amazed. You'd think that all of the baked goods in a prison would be shipped in from outside sources, but everything was made fresh on the premises, all of it run and operated by inmates.

Of all the places in the kitchen, the bakery is what really got my attention: the sweet smells, the sugary crusts on all the pies, the cloverleaf dinner rolls with butter seeping out of the creases. I'd always loved sweets. Growing up, we never had a cookie jar in my mom's kitchen. The only cookie jar in my family was at my grandparents' house—a green ceramic cat with big ears and its belly filled with the kosher cookies my granddaddy lifted from the Jewish bakeries in Westwood.

The next stop on my kitchen tour was the butcher shop. It looked almost like the hole because it was divided into a series of tight, caged-in rooms. Inside each cage was another, smaller cage that held the knives. I was, like, how are you going to give inmates knives to butcher meat? But it wasn't as simple as that.

A guard brought the inmate into a butcher cage and locked

him in. The inmate stood away from the rotating gate that separated the cutlery from the rest of the cage. The inmate then chose a knife from the guard's side of the gate, and the guard would place that knife on the gate and rotate it around to the inmate's side of the cage. Then the inmate would give the guard a chit for each knife he had selected, and the guard would place the chit in the case where each knife belonged. There was no way to get out of that cage without passing all the knives back through the gate first.

A couple of white inmates were at work in the cages. The guys in the butcher shop usually broke down turkey and chicken, but sometimes they'd dice up meat for beef Stroganoff, which I hated and they served at least three times a week. It was such a high-security area that I couldn't get close enough to the cages to see exactly what the inmates were doing, so I moved on.

The next kitchen was much smaller than the others, maybe the size of six cells. Another unusual thing about that kitchen was that it had a steel security door at the front. It was opened a crack, so I walked in and saw a little old white man sitting at a table. He had a home-style stove in there with a sink, a table, all kinds of cooking equipment, and his own personal walk-in fridge.

"What do you want?" the old man asked with a suspicious expression, as if I was going to jack him for some food or shake him down. I knew the brothers on the yard had a rep for putting pressure on the kitchen crews for extra food, so I knew what he was thinking. But I was just curious.

I said, "I'm new in the kitchen and I'm showing myself around. I had heard stories about the private kitchens and I just wanted to see it for myself."

"You must work the pot and pan detail," he said, as though he knew my number.

"What makes you say that?" I asked.

"I never seen you before in this area of the kitchen, and I know all of the blacks who run the hot line in the main kitchen."

"You're right," I said. "I scrub all of your dirty dishes and see all the food trimmings the rest of us don't get—all the things you guys eat."

"No one is allowed in this kitchen but the kitchen guards and the Jewish cooks," he said sharply.

I knew better than to talk shit to the old man. It wouldn't lead anywhere good. Besides, I wanted in on the kitchen hustle and the Jews had major influence with the warden. They also paid off some of the black shot callers with special food so no harm would come to them. I wanted to learn more, so I stayed calm and tried to engage him.

"So you are the man here?" I asked.

"I am," he said. "I'm the head inmate kosher cook." I kept talking and got him to feel important, bigheaded. Then he started giving me a rundown of the place, and the kosher meals he cooked for the Jewish inmates.

As he described his kitchen, all I could think about was how great the Jewish guys had it with their own kitchen and special meals. Then I noticed a stack of trays individually sealed with plastic wrap and asked, "You guys get TV dinners, too?"

"Those are for the Sabbath."

He explained how the Orthodox Jews observing the Sabbath couldn't work from sunset on Fridays through sunset on Saturdays, so all of those meals had to be prepared ahead of time.

The food in there smelled very good, so I asked him, "How does an inmate get to have kosher meals?"

"Well, first," he told me, "you have to be Jewish."

Every white guy, I soon learned, wanted to be a Jew—especially around the Jewish holidays. Every year at Passover and Rosh Hashanah, rabbis from L.A. would come in and do a whole big shindig. They'd have the entire dining room all to themselves. They had fresh potato pancakes, crisp salads, fruit platters, capons, brisket.... It was all top of the line, anything the rabbis could convince the federal government they needed to have for the ceremonies, they got.

The Muslims weren't far behind, either. When they broke their Ramadan fast at sunset each day, imams would come in and they'd have a feast. Then everyone wanted to be a Muslim. Every brother in the prison would be saying "As-Salaam-Alaikum" to one another.

I never took part in Ramadan because I couldn't fast from sunup to sunset for those thirty days. I was a hungry motherfucker. In prison I was making up for all the food I didn't get as a kid. I'd always tell myself, "Religion or not, I don't think the Lord will get mad at me just because I'm trying to eat."

After I got done talking to the head kosher cook, I made my way back to the main kitchen area. There were a few Mexicans and a couple of white boys in there, but the brothers had that kitchen pretty much locked up. Everyone appeared to get along, though. Later I'd learn that their good working relationship was based on the fact that it was a win-win for everyone. Aside from the daily leftovers, there were also samples from the big food service companies to be divided up.

Just like with a restaurant on the outside, all of the big food service companies wanted to do business with the federal Bureau of Prisons. With our facility serving up to six thousand meals a day, a CEO would have to be a fool not to try to get some of those federal dollars. So the companies would send samples of food into the prison to try to get their products placed on the menu.

Of course, a lot of that food never made it to the general population, so there was a freezer full of food that no one but the guards and the kitchen staff knew about. When the guards were in a good mood, they'd break out cases from the freezer full of items that most inmates hadn't tasted in years. We had steaks, shrimp, big pork chops, pizzas, frozen burritos, and cold cuts—all on the hush-hush.

The extras we got depended a lot on who the guards on duty were. There were usually four or five kitchen guards on at a time, mostly retired military men. From what I heard, they made more money than the COs. Most of them were cool. Officer P, he was a player from Long Beach. He was cool with most of the inmates. He never went overboard with the discipline. But sometimes we'd get stuck with these redneck or *house nigga* guards, as the black inmates called them, who played favorites with the kitchen workers.

Even in prison the white inmates got more love than the blacks and the Mexicans. There were guards on the yard with long ponytails, tatted-up, ex-military and biker types who ignored it when the Aryan boys broke the rules. The house nigga ones were always trying to prove themselves to the lieutenants by being even harder on the blacks than the white guards. We could only get a break from a handful of black guards, like P and Fish, who came from the hood.

Aside from the guards, the real underboss in the kitchen was Big Roy, a big fat black guy from the West Side of Las Vegas. He was the shot caller in the main kitchen. He ran the whole operation and even had influence with the white boys in the bakery and the Jewish cooks in the kosher kitchen. Before becoming a PCP cook, which was what turned him into an inmate, he'd been a sous-chef at the Horseshoe Casino in downtown Vegas.

I knew he was the one to get close to. Even though I was technically now part of the kitchen crew, I wasn't part of the cooking detail. The pot and pan crew was the lowest crew of them all. Even the guys who washed the eating utensils had it better than us. Nobody stayed on the pot and pan line for long. Either they moved up or they found a job somewhere else altogether. The older inmates were the only ones who didn't leave the dish crew. They actually liked it because there was no pressure, no competition, and they ate well.

Big Roy sweated so much that he had to strip down to his T-shirt in the kitchen, but he was a great cook and organizer. His food always reminded me of the flavors of my childhood, and being in my grandmother's kitchen. Big Roy put love into every dish, and breakfast, lunch, and dinner were always on time. Big Roy truly understood the importance of food to an incarcerated man. He gave us his heart and soul in the kitchen and knew that we loved the southern touches in his food. Even the warden knew about Big Roy's presence in the kitchen, and he had Roy oversee a special crew who cooked for the guards and the administrative staff in a private kitchen. Under the watchful eye of Officer P and another guard named Davis, he cooked them the same food that we ate, but used better ingredients, and they got more selections.

Roy's influence meant that he got first dibs on anything extra. Pastries, meat, chicken, fruits, and vegetables—nothing went to anyone else until it went through him.

This is how the hierarchy in most federal prison kitchens worked. There are always guards watching it all go down, and then there is always one powerful inmate cook, like Big Roy, who runs the kitchen, or there's a powerful crew of cooks who run it as a collective with part of the crew in every section of the kitchen. The head guy touches everything first, as soon as the guards pull it from the freezer or the sally port. Then he doles the food out to the prep cooks in the main hot kitchen and bakery, then the vegetable guys and the starch crew.

Next on the food chain are the servers, who get their cut from what is left on the hot buffet line after last call. The servers have the most dangerous job in the kitchen, because if you had just one small piece of chicken left and some gangster wanted a bigger piece, you had a real problem; even though it's not a server's fault when the food runs out or if there are only a few small pieces of chicken left, a lot of prisoners still go ahead and kill the messenger—not necessarily literally, but most will start a beef with him.

The last crew to get its cut was the dishwashers. We got the last pick of everything, but it was more of a barter system than just a handoff down a line. As with everything, it was all about leverage: We had the least and Big Roy had the most.

Big Roy ran the meat crew, seasoning and preparing the beef, chicken, fish, and stews. Once the food was cooked, Big Roy made sure to cut a share of the hot food for the white boys running the bakery in exchange for his share of the rolls and sweets. The kosher dudes got kicked down next, because they had what no other kitchen had access to. Their packaged kosher TV dinners were easy to smuggle back to the units, and those kosher Sabbath dinners were always a hot item. The chicken meals could fetch $10.00 a pop, and the kosher cooks always made a killing on what the rabbis brought in for the holidays.

Whatever Big Roy didn't eat himself or hand down to his crew or trade, he sold. He was really in cool with the white boys and the Jews when it came to that business, but he didn't like dealing with the brothers because they'd always try to strong-arm him for cheaper prices. The black guys didn't mind paying two bucks for a chicken breast and a wing, or a thigh and a leg, but Big Roy could get double that from the whites. The brothers knew they were getting cut short, though, and from time to time someone would want to stick Big Roy. So, Roy had to kick down some of his own stuff to certain brothers on the yard—the shot callers—to keep himself protected.

Roy's best customers were the older, wealthier white inmates (the mobsters and Wall Street moguls). They hardly ever showed up at the mess hall. Instead, Big Roy served them right in their cells.

Everyone knew Big Roy and everyone wanted something from him. As far as inmates go, he had a lot of power. From the first day on pots and pans, I knew what I wanted. I was never cool with being small-time—that's what got me locked up in the first place: I wanted to be the man. Now, I knew who was the man in prison. Like T-Row before him, I wanted to walk in Big Roy's shoes.

But after a month, it was starting to look like there would never be a chance for me to move up. Cooking opportunities were few and it took time to get one. I kept having to remind myself that I had almost another twenty years—it helped me learn patience, but fucked with my head.

My chance to cook for the first time came a few months later. It was Juneteenth, an African American holiday that celebrates the freedom of slaves, and the Black Culture Workshop, a black self-help prison organization, got the warden to approve a special menu in observance. I had stayed behind several days in a row to help the black cooks prepare a soul food feast of fried chicken, catfish, dirty rice, collard greens, corn, and peach cobbler.

I knew that they'd need a lot of help on the day of the feast, so I decided to skip my workout session and asked Officer P for extra duty. Big Roy was walking around the prep kitchen driving the crew, tasting everything and schooling the crew on what he wanted. There was more work than they could possibly handle, and so I jumped right in. With no real experience, I grabbed one of the large paddles and began stirring the collard greens that were in a really big kettle. Big Roy had them simmering with neck bones, onions, and bay leaf.

"Jeff, if you want to get down with us," Big Roy bellowed from behind me, "I need you to wash and prep the chicken."

I was happy as a motherfucker to hear that. Just to be in the kitchen, to be part of Big Roy's crew was all I wanted. That day I washed and seasoned more than two thousand pieces of yard bird, and I did it as quickly as I could. The whole crew seemed impressed with my drive, but still there wasn't a regular spot for me.

For weeks after that day, I kept after Big Roy, and he'd tell me, "I don't know, youngster, you never know what's going to go down in this place. A lot of people have been waiting to get on the cooking crew."

I kept volunteering for extra duty, hoping to get off pots and pans. A month and a half later, some of Roy's own crew got put in the hole for having dirties, negative drug tests. His response changed: "Youngster," he said, "if you still want to get down in the kitchen with me, this is your chance."

After busting my ass in the kitchen, proving myself on pots and pans, and volunteering whenever possible, I finally got my break. I was issued my kitchen whites—a short-sleeve V-neck shirt, white stretch cook pants, and black high-top nonslip shoes. We weren't given toques; we wore old-school paper diner hats.

I didn't know shit about mass production cooking, or restaurant cooking, or any noncrack cooking! The only things I knew about working in a kitchen were the things I picked up from watching Roy's crew since my first day on pots and pans. But I was anxious to learn and to prove myself. Right from the start, I banged hard for Roy.

The kitchen itself was big, with little windows that faced the Coast Guard station. All of the equipment was chrome plated, and the prep tables were shiny steel. My first assignments were to learn where everything was kept and to get things the other cooks needed as soon as they asked. Big Roy and the crew had me doing all the bitch work, but I was hungry to be part of the crew.

Once I got to do actual cooking, it didn't come easy: I screwed up the vegetables a few times by boiling them too long, or forgot to add salt to the water, or didn't have an ice bath ready to shock them (plunging them in ice stops the cooking once they come out of the boiling water). When I overseasoned the meats or burned things, Roy would rough me up about paying attention.

Still there was something about me that Roy liked. He would often pull me aside and show me how to do certain things, take his own time to train me. Because of these talks Roy gave me on cooking techniques, some of the other guys started hating on me. He wasn't nearly as patient with a lot of the other cooks. I think it was because it was clear I was prepared to do whatever it took to make it in his kitchen.

The competition in the kitchen was intense at times. But the pressure while cooking was nothing compared to the constant presence of danger as I walked the serving line, restocking the two hundred pans of food for the servers. Fifteen hundred convicts came through that line three times a day. These guys would stare you down hard. You had to be strong or at least look strong. In the summer the swamp coolers would often fail in the kitchen and the heat drove tensions even higher, especially between inmates, guards, and cooks trying to serve the food. Fights sometimes broke out just over the size of portions served.

There were times when I wasn't sure that the kitchen was for me. I wasn't ready to die over a piece of fucking chicken, or get beat down because I chose to sell my share of bananas at a marked-up price. Sometimes focusing on the job was a challenge. I burned my hands on oven doors and cut myself on the number 10 can lids. (I didn't know it then, but I was learning a lot of bad, even dangerous, cooking habits in that prison kitchen.)

While prepping, I often drifted off thinking about times in my life when I was a free man. Cooking took me back to the Motel 6 in San Diego. I was at the stove cooking pounds of cocaine and watching it harden as I submerged the glass pots in the ice-cold water one at a time. Then I'd scrape the stove for every crumb of residue and recook it. Was I the only one thinking this while plunging frozen vegetables into a large kettle of boiling water?

The more I thought about it, the more my past was beating me down. I was sweating my ass off among the dregs of the world, but it wasn't just my fall that I was thinking of. It was that I had descended to the lowest place a person could fall and that, as far as America was concerned, I was exactly where I belonged—locked down with the scum of society. I wasn't recalling the good old days of my ill-gotten gains, I was regretting them. As the steam burned and pruned my skin, as I looked around at my fellow inmates and compared this to my life as a high roller, I finally knew this was what America thought of me. I was just a petty criminal. And worse. All the wrong shit I did my whole life started to become painful. The kitchen made me face it head-on. It stopped me from pretending that I did nothing. I could no longer hide from it or ignore it. I had to move on, and eventually these thoughts drove me to want to be in the kitchen all the time.

Throughout the prison all of the inmates had created these little communities, in part to try to keep on being the person they were on the outside. Me and my homies were exactly that: the same fools we were on the streets. We spent most of our time talking about the money we had had, the cars, the women. . . . Some of them were just waiting to get out and would probably be back inside one day again—or dead. I was no different. If I had gotten out then, I probably would have been back to my old ways. But I was now starting to see that maybe there was something more I could do with my life. I began to dream of a better life. My homies in the other pens started seeing less of me as I began spending more time in the kitchen and studying the world outside of Terminal Island.

I spent more of my time watching Big Roy and all of the other cooks. I thought about cooking all the time. I even wrote down some of Big Roy's recipes and looked them over at night when I was in my cell. By the light from the small lamp I had, I committed those recipes to memory and went over each step again and again.

I was learning to cook and was proud of how quickly it came to me. Enough that I started sharing my cooking experiences with my family, telling them how I'd cook for them when I got out, how I'd one day have my own restaurant.

TEN

～～～

DAYLIGHT

As much as I enjoyed my kitchen duties and all the extra food, cooking didn't pay as much as some of the other prison jobs, and my money was starting to get short. After my arrest, I had my father collect all the cash from my various stashes—maybe $60,000 in all—plus my diamond-encrusted Rolex, my gold rope chains, and diamond rings. Dad was the only person I trusted to keep my loot. Not that I didn't trust my mother, but I spent so much money on her over the years that I was afraid she might have developed a taste for luxuries that would tempt her to spend my money before I ever saw freedom again. I didn't want Dad to keep putting my savings on my prison account because I knew I'd need it when I got out.

My commissary shopping list was pretty consistent. Every Thursday, I purchased two bags of Doritos, four or five avocados, Top Ramen noodles, cans of tuna fish and chicken, Kern fruit juices, some photo tickets so that I could take pictures on the yard and in the visiting room, and a few stamps and greeting cards to send to my family, some old friends, and girlfriends.

I'd never been in love as much as when I was in prison. I loved everybody. Out on the street, I tried to pull every woman I could, but prison brings out the more affectionate, emotional side of a man. And, after Dana left me, a lot of women from my past got in contact with me. Girls from as far back as high school started writing me. Tammy still visited once in a while, but she only let me see my son if she brought him herself. It was her way of punishing me because my mother once brought him for a visit and then let Dana take him home because Moms had to leave early.

I always purchased Ramen noodles at the commissary. I hated Ramen noodles, but I'd use the seasoning pouches to add beef or chicken flavor to the tortilla chips for the nacho parties we threw three or four nights a week. We all smuggled something different out of the kitchen, or bought it from Big Roy on the yard. We got things like cooked hamburger patties, cheese, diced onions, canned jalapeños, so we could feast while watching *Midnight Love*.

Everyone always asked me to put together the nachos since I had a knack for it. Sometimes I'd take canned tuna or canned chicken and add that to the nachos and throw it all in the microwave (which was kept in the common area where a guard could monitor its usage). That was always a real treat.

Even in prison, Christmas was a special time of the year. For the holidays, we were allowed to order those Pepperidge Farm variety packs, though each inmate was only allowed five of them because the Feds didn't want us stockpiling food in our cells. They had Gouda and all these other fancy cheeses I'd never eaten in my life, and they sure made our nachos taste different. We'd take the sausages from the packages, slice them up, cook them full blast in the microwave until they were fried, and use them to top off the nachos.

About a year after I first started in the kitchen, my caseworker, Mr. Welsh, came around and told me I had a visitor. Now, that didn't make sense. It was a Tuesday morning and we didn't have morning visits during the week. So right away I knew something was wrong, but Mr. Welsh wouldn't tell me what it was.

I became even more nervous when Welsh handed me over to a Special Investigative Services guard. I couldn't imagine what they thought they had on me. The SIS guard took me to a private room where I came face-to-face, once again, with the narcotics agents from San Diego and L.A. There was another man in the room as well. Turned out he was an IRS agent from Los Angeles. He told me to sit down.

"I know you're doing twenty years," he told me. I just shrugged my shoulders—like that was news. After a few minutes of small talk, he told me they were investigating some guys I once dealt with and were putting together a list of potential witnesses in case they went to trial. I wasn't surprised to find out who it was they were trying to make a case against. I mean, every major crack dealer on the West Coast and in the dirty South had dealt with these guys in the 1980s on some level and they never caught a single case that I knew of.

I'd been locked up for three years and almost every man inside was suspected of being an informant. Just about everyone I dealt with in L.A. and San Diego was suspected of being hot. Before I caught my case, people thought I was hot myself because I'd never been busted. The Feds were notorious for putting snitch jackets on those who wouldn't play ball by spreading false rumors, implicating them by putting their names on government paperwork.

I thought about the speech my sister, Cali Slim, gave me when the government offered me a plea bargain just before my trial. She said, "Save yourself, Jeffrey. These guys don't give a fuck about you. Not one of your so-called homies have come to see you or helped your family. I've seen you help every one of them again and again, proving your loyalty again and again. Those big-time dope dealers got rich off you and left you for dead when you got busted. You got Sweet an attorney; you did everything you could to save him from the Feds."

I was in a trance listening to her break it down to me, but I told myself I couldn't go down like that. I had a rep, was loyal to the game, and never wanted to wear a snitch jacket. I was a proud hustler. At the end of day, that's why I got twenty years. Being in for three years and looking at another fifteen, I was listening a little more now.

The IRS agent told me the guys they wanted to bring down had dealt with so many people that they'd never know who was on the list of potential witnesses, that they'd already interviewed more than thirty people.

These particular guys had a legitimate business to front their drug and money-laundering operation, and the agent told me all I would have to do is agree to testify that I'd done both legal and illegal business with them. In return, my sentence would be reduced to the ten years and seven months they initially offered me before my trial.

I asked the IRS agent, "How'd you know I had been dealing with them? That was years ago."

"We have people who told us, plus documents," the agent said. "And it's someone you worked with who was very loyal to you. This person isn't under any indictments at all. This person doesn't need to say a thing. This person was willing to help us if we were willing to help you—and if you were willing to help yourself. That's why we're here."

I thought about that. I'd been true to the game, done my trial, taken my nineteen and a half years, and niggas were *still* running their mouths about me, but someone out there was trying to help me, force me to help myself.

I asked them who it was that put my name on the list, and then they said the last name I would have expected. She had embraced the game as much as I ever had. She had been right there with me, my partner in crime, Bonnie to my Clyde. Then she betrayed me nearly the minute I was off the streets—running with other guys, not answering my calls when I needed her the most. As far as I knew, she was still living the high life with some other hustlers, but it was Carmen who was trying to help me now. Reaching out from wherever she was to give me back some of my life.

I told them I'd think about it.

No one with a major federal case can tell me that they never thought about *talking*. With the Feds, it's always you or the next man.

Two weeks after that meeting, I agreed to let the Feds put my name on the list, but my mind was still very much that of a street hustler, so I had to justify making the deal to myself. I kept telling myself that if I did have to testify, I would back out. I wouldn't trade my freedom for my family's safety. To this day, my decision is still a burden on my conscious. Jamar, my son, was growing up, my parents were getting older, and my grandmother was near death. I was working hard to change my life, and I knew that if I ever got an opportunity to be free again, I'd spend the rest of it trying to give back to those I'd taken from. Still, I worried what people would think about me if I told.

I still considered myself a hustler and was loyal to the game. It was all I knew. What would T think about me if rolled on these guys? What would all the homies think? I had too much pride to do it, but that same pride had gotten me where I was, looking at two decades in a cell.

"Fuck them," Cali Slim told me on the phone. "Those mother-fuckers don't give a shit about you. You do what you've got to. Stop being hardheaded. They showed their love for you by abandoning you. No lawyer, no letters, no money. They didn't even look out for your son. No bikes for Christmas, no clothes for his birthdays like they promised."

Then she brought up something I'd been trying to block out of my mind ever since I first went to jail. "You weren't locked up for forty-eight hours before your so-called homies broke into your house and stole all your TVs and stereos. Even after that, they *still* ran your name in the gutter, hating on you. You came into this world by yourself, Jeffrey, and you gonna leave by yourself. You better put your pride in your pocket and forget about that homeboy hustler shit and be your own man."

I knew at this point in the game that I would have to make a conscious decision about my future. My family, especially my son, became my number one priority. Not the homies, not the females, but directing my son Jamar down the path of success. I'd let the past be just what it is: the past.

Having made that decision, I intentionally set myself even further apart from the homies in prison with me. Partly because it was too hard for me to face them on the yard. I felt guilty and weak, even though I had never betrayed them. And partly it was because I knew I had to get away from them. I had made the right choice. I would have another chance at a real life, and that meant not going back to who I was.

During this time, several white boys in my unit took an interest in me. I'd heard they were big-money guys. They were always reading the papers and watching CNN in the white-boy TV room. What really caught my interest was *60 Minutes* and *20/20*. I used to watch them through the window to see what they had on the TV, and even sometimes exchanged a few words. Eventually they invited me to come in and sit with them.

One night one of the older white dudes said to me, "Henderson, you seem pretty smart, but I notice you hang around with some of them thugs on the yard. Somehow I don't quite see you in that life."

"Is that right?" I asked him. "What do you mean?

"You seem intelligent. I see you in the law library a lot, and I see you reading over by the Birdman's spot on the North Yard practically every day."

"Yeah, well, I haven't always been a reader," I replied. Now education is important to me, I told him. I had just graduated from the G.E.D. program. It was a big moment for me. My father came to the graduation; we ate food and took photos. "Since then," I continued, "I've been trying to stay focused on becoming a chef. That's kind of my dream." It was more than that, though. As I'd gotten better as a cook, I'd started reading books about how to become a chef on the outside. I was developing a blueprint for life when I got out.

"Well, you're always welcome in the TV room," he told me, and he meant it.

After that, the white-boy TV room became my regular hangout. I got exposed to current events, learned about how the government ran, how the media worked. I even began to understand how the different political biases affect the news.

The white guys used to always be on the North Yard when the newspapers came in the morning. Only fifty papers came to the prison each day. The white boys would sit back while the brothers jumped the line, dissing them to get all the papers first. But most of the brothers only took the sports sections and threw the rest aside, so the white guys would just stroll over and take what they wanted after the brothers dispersed.

I read the *New York Times, LA Times, Wall Street Journal,* and *USA Today* and got a whole different perspective than I ever got from Kevin X or Imze. Those white boys weren't just intelligent; most of them didn't fit the profile I'd been reading about in all those black empowerment books. They were individuals; they spoke about their friends and families on the outside, reminisced about the good times gone by. They sure didn't strike me as wicked men plotting to keep the black man down. They seemed cool to me—caring and unbiased. Unlike anyone else I'd ever been around, they told me I was smart. Maybe they were just trying to flatter me, but they were the first people who ever said anything like that to me.

Their encouraging words helped me to believe in myself and to have confidence that I could make something of myself upon my release. It was still many years off, but life after prison now seemed like a real possibility. I had given up on everything, on life; now I was planning for it.

Two weeks after I put my name on the list, I was awakened at the usual 4:00 A.M. and brought to the receiving area where I had first been processed into Terminal Island. They told me to change out of my khakis and issued me a BOP jumpsuit. Myself and four or five other inmates were loaded into a van and taken downtown to the FDC Los Angeles. Everybody was silent for the ride. We didn't know one another's business, but there were only three things we could be going to do: testifying, getting resentenced, or being charged on a new case.

At the FDC, they put us in a holding tank with a bunch of other inmates. One guy was Harry-O, who allegedly cofounded Death Row Records while serving twenty-five years on a murder rap. He didn't know me, but I knew him.

I sat in the tank all day without ever being called. Apparently,

I was on standby in case these guys went to trial. But the guys they were making the case against decided to plead out, take a deal, and snitch out their ties to organized crime.

Around 4:30 I was transported back to Terminal Island. I was feeling fucked up the whole day. My conscience was messing with me. Back at the prison, I'd have to face all the homies and the brotherhood, who'd be like, "Why'd you go to court? You got nineteen and a half years."

I told them that the government was trying to trump up some more charges from back in the day. Everyone let it be—hell, most of them had no grounds to question anyone—but before long, rumors were circulating around the yard.

Behind my back, brothers were whispering, "I bet he went down there and told on somebody."

I didn't feel bad. For one thing, I could honestly say that I never testified against anyone, and no one was in prison behind me. Nobody I ever fucked with on the streets was in the system behind Jeff. I was willing to take that to my grave.

Three months had gone by when I got a letter from my attorney informing me that my sentence had been cut in half to ten years, seven months. I immediately went to my caseworker, Mr. Welsh, and asked if I was now qualified to have my security level dropped and be sent to a prison camp. I'd always heard good things about those prison camps—the "Club Feds." They had no fences, the food was better, and there was very little tension. Mr. Welsh said he would see what he could do.

About six months later, in the summer of 1992, I was notified that I had been accepted to Federal Prison Camp Nellis, at Nellis Air Force Base in Las Vegas. I told no one about my sentence reduction and that I would be leaving soon.

As I walked the track the night before I was shipped out, I thanked God for delivering me to a point in my life where I saw hope and could feel freedom. I would miss the brotherhood and the corporate white boys—both had prepped me for change and showed me the way to my soul. I had gained a sense of who I was, had learned to deal with what I'd done, and had even found hope. And my time in the kitchen had given me direction. For the first time, I knew what I wanted to do with my life.

ELEVEN

‿‿‿‿

MEAL TICKET

In the summer of 1992, after another 4:00 A.M. wake-up, I was
back on the chain gang in the white BOP van, heading for desti-
nation number three. It had been three and a half years since my
arrest. I knew the trip to Las Vegas would be a long one, and I had
four hours to savor all the sights of freedom on the road: the new
cars, buildings, and restaurants, including my favorite, Popeye's
chicken. Through most of the ride, though, I just reflected on my
past and started thinking about my future, now that I was able to
see daylight at the end of the tunnel.

As we approached Vegas on Interstate 15, I saw all the lights on
the strip coming on. *Wow,* I thought, and felt free for a moment.
But then the marshal's two-way-radio chatter kicked in again and I
was brought back to the reality of where I was headed.

Reaching the air force base, my stomach began to knot up like
it always did when I arrived at a new prison, not knowing what I
was facing. I told myself, "Hell, it's only Club Fed."

Once we were unshackled and let into the camp's inmate receiving area, I realized that this would be nothing like my previous
experiences with incarceration. There was no fanfare, no ocean

of shouting faces. It looked like what I imagined a small military base would look like, complete with rows of barracks and carefully manicured lawns. Everywhere I looked, I saw inmates in khakis, airmen, BOP guards, and civilians all intermingling.

After the usual round of questioning, I was issued a dorm unit, given a new bedroll, and sent on my way. I knew some homies from the 619 were there, so I sought them out. I wasn't going to get mixed up in any nonsense, but just the same, I was new to the camp and I needed the rundown. I found my old friend Chris Wright in my dorm. He was doing time with three of his brothers. He gave me the standard care package that all inmates received, showed me the phones, and walked me to my bunk area. When I was done unpacking and setting up my bunk, it was nearing count time, so Chris gave me a quick tour of the camp.

I was in awe. There were no fences, no barbed wire, no gun towers. And there was no stench of tension between gangs or races. The place had basketball courts, indoor racquetball, a baseball diamond, and a gym in the basement. It even had carpeted dorm rooms and cable TV with a few premium channels. I soon learned that despite all these great amenities at Club Fed—things you never get in medium- or maximum-level federal prisons—doing time was still time. The physical aspects were better, but this was an even tougher mental war. Here you could taste freedom; it was right in front of you, but you couldn't touch it. You couldn't cross the line, because you'd lose all the comforts of the camp and head back to a higher-security place facing even more time than you had before.

I was assigned to an eight-man dorm. After count, I got acquainted with my dormmates. Everyone was pretty welcoming, offering me cosmetics, and chips and sodas. And just like at Terminal Island, the guys were doing nacho feasts. I was right at home and felt relaxed for the first time in years. Relaxation soon turned into urgency.

All I could think about every day was getting to see my woman—I had gotten close with one of my old girls Shawn while still at Terminal Island and she had promised to visit me at Nellis, where we might actually get time alone—and getting a gig in the kitchen. The camp's mess hall was just like a restaurant. It was called the Red Horse Dining Hall. From the first time I went in there and saw the upscale prison food that was being served and watched the inmates working the hot line alongside the military personnel, I was eager to be a part of it.

My first job at Nellis was to clean the bathrooms. I didn't mind doing the work; it was just like being back in the day with Granddaddy. I was an old pro at cleaning toilets; I still had the Tidy Bowl touch, but I became frustrated after several weeks. All I thought about was how I could get into that kitchen.

Where I chose to spend my nights during my first few months at Nellis helped me get into the kitchen during the day. I hung out in the band room after walking the track in the evenings and listened to the musicians. I especially liked listening to their bass player, Friendly Womack, the eldest brother of Bobby Womack, who had been a member of the 1960s gospel band The Valentinos. He was serving a five-year stint for trafficking and was also running the inmate kitchen crew. After I listened to him play, we talked a lot about cooking and I told him how I was going to make something out of cooking when I got out of prison. He shared the same dream and we quickly became friends.

Friendly was the chief inmate cook at the Red Horse. As soon as I felt comfortable enough, I asked him to help get me in the kitchen. He had me write a job request to my caseworker and assured me that he'd put in a good word. He even helped me fill out the inmate request form and he introduced me to one of the cool air force sergeants named Norma. After several weeks of waiting, I was in the kitchen. But this time, having a little experience from my kitchen stint at Terminal Island, I started near the top, as an assistant cook. Friendly liked my enthusiasm and that I was a quick learner.

Friendly reminded me of Big Roy. Both of them had a great palate, demonstrated strong leadership, were very talented at seasoning food, and were organized in the kitchen. The only real difference between the two of them was that Friendly took care of everyone, including the brothers without much money. And Friendly was real messy. Within an hour, the kitchen had food all over the place.

The first thing Friendly taught me to bake was buttermilk biscuits. They weren't made from scratch, but at the time I didn't know there was a difference. There was a bag of premixed batter and all you had to do was mix in buttermilk and chunks of butter, and then bake them. Every morning at 4:30, Friendly and I would be up getting the biscuits in the oven first thing. Next was the gravy. We started by sweating diced onions and bell peppers, and then added whatever leftover meat was around—from turkey to ground beef. Once the meat had some color on it, we drained the fat and seasoned it with salt, pepper, garlic powder, and onion powder. In a separate pan, we made a roux with flour and butter, then added some heavy cream and stirred in the meat. We let it simmer all morning, and then poured the gravy over the hot biscuits. Military lingo for the dish is "shit on a shingle," but it tasted great.

Next, he had me make the pancakes, stacks and stacks of pancakes. Then he trained me on the line cooking omelets. The military folks used to come in along with the inmates, and the made-to-order omelets were one of the most popular breakfast dishes, so I had to get good quickly. I used to have fifteen or twenty omelets on the flattop at one time. I took a pitcher of prebeaten eggs, poured some on the griddle, and worked them up with my metal spatula—adding cheese, mushrooms, bell peppers, whatever filling was ordered. As soon as I served up an omelct, I slapped the counter with my big spatula and called out, "Next! What can I get for you, sir?" I made that griddle sing.

Friendly soon paired me up with Big Joe on pantry detail. In my mind at that time, Big Joe was a salad master. He taught me how to prepare Thousand Island and blue cheese dressings from scratch, even though the military provided him with powdered mixes.

After my salad training, Friendly broke me in on the lunch menu, where I learned to make lasagna, meat loaf, meatballs, smothered pork chops, and, my favorite, fried chicken. Friendly instructed me on how to operate some of the heavy equipment. The Hobart mixer, the braiser—all the machines I'd seen at Terminal Island, I was finally learning to use myself.

After a couple months, my confidence was riding really high and I felt like a real cook. I owed a lot of my success to Friendly, who I learned was soon being released. The head inmate cook job would open up when he left, and he had been prepping me to be ready for it and pushing me to go for it.

Friendly and I became so close that he finally shared his secret recipe for his famous jailhouse fried chicken with me. All my life, fried chicken had been my favorite food—I ate it every chance I could, but I never knew how to fry it myself. When Friendly gave me his recipe, I thought I was the man.

I don't make a lot of fried chicken these days, but when I

do, I still use the recipe for Friendly's Famous Buttermilk Fried Chicken. I would put this recipe up against Popeye's any day:

2 tablespoons cayenne pepper

3 tablespoons black pepper

4 tablespoons kosher salt

2 teaspoons onion powder

2 teaspoons garlic powder

2 cups all-purpose flour

1 chicken, cut into eight pieces

quart buttermilk

1. Mix all of the spices together in a bowl. Put half the seasoning mix in another bowl. Add the flour to one bowl, mix well and set aside.

2. Rub the chicken with the reserved spice mix. Poke all of the pieces with a fork a few times and set aside. (Friendly taught me to pierce the chicken pieces with a fork so the buttermilk seeps down into the bird.)

3. Pour the buttermilk into a stainless steel bowl. Add spice mix and the chicken pieces. Cover the bowl with plastic wrap and refrigerate for an hour.

4. Dip the chicken pieces into the seasoned flour, pat the pieces together and make sure they are heavily coated.

5. Drop them into a deep fryer, like we did, or in a deep pan with enough vegetable oil to cover the chicken. Turn the chicken as it browns and remove once done. Soon you'll be having Friendly Fried Buttermilk Chicken.

We were the talk of the camp. Friendly had the trust of the military, the guards, everyone at the prison camp. Me, I was still something of a hustler. I decided I could have the same operation Big Roy had going back at Terminal Island, and I convinced Friendly to go along with this. He would fry extra chicken and I would smuggle it out to the unit. I was just like a one-man Popeye's, selling two-piece, four-piece, six-piece, and fifteen-piece meals. Unlike Big Roy, I made sure the brothers got theirs first. Besides, most of the white boys didn't want their chicken fried, so we made them a separate batch that was baked.

As much as the kitchen was the centerpiece of my prison life, I found something else just as rewarding. I was asked to join the inmate T.A.P. (Teenage Awareness Program) by the prison captain. Along with a few other inmates, we were escorted to local high schools in Las Vegas to talk to at-risk kids about the decisions they were making. We tried to tell them about making the right choices, mostly though we ended up telling them what prison was like. Speaking to the next generation of young people who were headed down the same path I went gave me a sense of importance. I felt the power of my past working to undo my wrong, and that helped drive me forward.

At the camp, my daily routine was to be in the kitchen at 4:30 in the morning and out by 12:30. Then I sold my chicken and hit the library to read self-help and culinary books that I had been accumulating since I first got interested in the kitchen on Terminal Island. Around 3:00, I hit the gym, and then went to 5:00 P.M. chow. At night, I watched Friendly in the band room, and then hit the track for two or three hours, talking to some of the homies about what we were all planning to do once we hit the streets.

Friendly was three months short and wanted to make sure I could take over the kitchen when he got released. He took me through all the recipes and the overall kitchen operation again and again until he was sure I had it down tight. He even made sure that I had a good relationship with Staff Sergeant Norma, who was in charge of the Red Horse. She was a cool sister from New York City. She knew the chicken hustle we were running, but she looked the other way because our cooking and organization made her look good to her commander—besides, she knew the importance of food to inmates.

The day Friendly was released was a sad one for me. He was my best friend in there. I felt I was ready to take over the kitchen, but I was still nervous about it. Friendly had watched over my every move, and he had schooled me a lot about life. Working without his guidance was going to be the biggest challenge I'd faced so far. I was to be the top dog in the kitchen. The time had come for me to utilize the cooking skills I'd learned from Big Roy and Friendly, and my talents for being a natural leader that had helped me excel on the streets.

I saw Friendly out to the control center and watched him get into the BOP van and head off toward the airport. Now the whole kitchen was on my shoulders.

After my first week running the kitchen, I started to change a few things. Friendly had never let me tweak his recipes, but now I saw an opportunity to make a name for myself. I put together my own kitchen crew. As the inmates left who didn't want to roll with me, I replaced them with my handpicked guys. I trained them the same way as I was taught. As the leader, I was a control freak. I wanted to know every detail of what my crew was doing. I wouldn't let anyone put their own twist on my food. Prepare it like I taught you, I told them. I quickly earned the respect of my peers, and as they got with my program, I cut them into my chicken hustle. My kitchen routine ran like clockwork, and I delegated tasks like a four-star general.

The praise that started coming in from both the inmates and the airmen strengthened my self-esteem in a way I never felt before. In a strange way, it reminded me how proud I'd been on the streets when some crackhead would compliment me on my precious rocks, how I once called myself a "gemologist" whenever people asked what I did for a living. But now I was being praised for giving people something good, something they needed. As many times as I said it to myself before, now I knew that it was true: I would never be in the cocaine business again.

My caseworker noticed the progress I was making as a cook and told me about a culinary program for inmates in Maxwell, Alabama. She asked if I was interested in taking the course and working toward getting an associate degree in culinary arts.

"Definitely," I told her. I knew that a degree would give me an edge once I was back on the streets. I hated the thought of leaving my job at Nellis, but this was my chance to get ahead when I was a free man.

A few months later, I was approved for Maxwell. I was given a one-way ticket to Alabama, and they dropped me at the airport. With the Feds footing the bill for my flight, I'd be making layovers in several states and for the first time in a long time, I'd be doing it alone. I was eligible for this furlough because of my new status as a minimum-security prisoner and my history as a nonviolent offender. I had to sign a waiver before I started off, which explained that the Feds would slap an additional five years on my sentence if I didn't show up at Maxwell on time. This was a program that the Feds set up at all of their minimum-security prisons, mostly to save money.

I was nervous the whole trip, and uncomfortable around so many *regular* people. Being strapped into my seat, I was tense and uptight. Everyone around me seemed to be relaxed and friendly, but I had the most serious demeanor. I couldn't help it. I tried to calm myself, to be like everyone else, but I couldn't. To some degree I was institutionalized. I came to realize on that trip that when I finally got a taste of real freedom, it was going to be a struggle.

Even though I had no guards traveling with me, I felt more imprisoned than if I did. The temptation to run away was with me every step I took, but so was the thought that someone was watching my every step. I had eighteen hours to report to my new prison; I made it to the Alabama camp two hours early.

After the usual round of searches and interviews, I was escorted to my unit, which was a lot different than Nellis. Instead of eight-man rooms, the camp had two-story brick dorms with row after row of bunk beds. I didn't like that at all—too many asses and feet. The Nellis setup made you feel like you were in a little family. I'd also heard that these southern guys were two-bit hustlers and mostly snitches from the East Coast, but I tried to ignore all that and keep my focus strictly on getting a job in the kitchen and earning my culinary degree.

The next morning I arranged a meeting with the BOP chef foreman and asked about a kitchen job. Maxwell wasn't anything like Nellis. At Maxwell, white boys had a lock on the kitchen, and brothers from the Nation of Islam ran the bakery, and both groups were very cliquish. The best I could land was a server position on the food line, which would give me an occasional opportunity to cook and get extra food to eat. I thought it would at least get me in the door and let me work my way up to be a cook, and possibly head cook. Within a month of my arrival, I had an argument with the BOP chef about how to make lasagna properly, and I was thrown out of the kitchen and into a job on the golf course, tending to the fairways.

At least I had the culinary program to look forward to, or so I thought. Three days before it was set to start, the prison canceled the whole damn culinary program indefinitely. It was rumored that one of the female instructors and an inmate were caught having sex in one of the kitchen's walk-in refrigerators. My transfer had been a mistake, and I was really depressed.

It took six months of pleading my case to the warden before he finally transferred me back to Nellis, where I had to take a job picking up trash on base because a whole new crew had taken over my old kitchen.

It looked hopeless that I'd ever get back into the kitchen at Nellis, and I was growing more and more anxious to get out of the pen. Then I heard about a new drug treatment program. Even though I never used drugs and hadn't even been around any since I stopped selling them, the word was if you passed the program, you'd get a year off your sentence. Like a lot of inmates, I wanted into that program.

While I waited, the time passed more slowly than ever since I started doing minimum-security time. I was finally accepted into a nine-month drug treatment program in Oregon for first-time offenders who had no history of violence and no gang affiliation. I'd never even used drugs, just hit a joint once or twice with T, so I wouldn't look soft. Completing the program would get me out earlier, and it would mean spending the last six months in a halfway house instead of prison.

The two-hour flight from Las Vegas to Sheridan, Oregon, felt like it took all day. Now that I was in the program, I had exactly one year left to serve. If all went according to plan, I would be walking out at thirty-one—fresh, intelligent, invigorated, redeemed, and ready to take on the food world. I was thinking about my future all through the flight. I knew I didn't want to go back to San Diego, where the old homies might pressure me to hustle with them. I would head to L.A. and start fresh. My grandparents were getting on in years and I figured I could stay with them when I got out of the halfway house.

Thinking back on the seven and a half years I'd been locked up, I realized that I hadn't been arrested, but that I'd been rescued. I found out who I was in prison and became a man. Kevin X and Imze taught me a lot about pride, knowledge of self, and black people's history in America. And those white boys in the TV room gave me a broader perspective to build upon.

"It's up to you to find your own way," one of them told me. "You can't get your blueprint for life from another man." I'd never forget them telling me that I was smart, different. I'd never forget all the news programs and newspapers that took me out of prison and brought the world to me.

The drive to the camp seemed as long as the flight. There was more wilderness than I had ever seen before, and nothing but a dim, gray sky everywhere I looked. I didn't care if it was cold and rainy all the time, since I was twelve months and a wake-up to the halfway house. Doing time was a science. I did the time; I didn't let it do me. As long as I had reading material, a book of stamps, and a kitchen, I could do another year standing on my head.

After the usual third degree in inmate receiving, I was escorted to the RDAP (Residential Drug Abuse Program) unit. All of the RDAP participants lived in the same dorm, so that we could consult and support one another through the course of the program. I didn't give a damn where I stayed, so long as I got through the program and got home.

Once again, I asked my caseworker for a kitchen job; once again, they were all locked down. And, once again, I wasn't going to just take no for an answer.

The food at Sheridan was pretty good, but what really got my attention were the baked goods. Whoever was doing the baking had it down to a science. I had already mastered prison cuisine at Nellis, so I decided it was time to learn something else. The kitchen foreman was a cool guy named Mr. Pane. I bothered the poor guy about getting me a job in the bakery for three months before he finally relented and gave in.

In the meantime, I had RDAP to focus on. There were only about fifteen of us in the class, but we represented almost as many religions and ethnicities: blacks, Jews, Christians, Nation Muslims, Sunni Muslims, Indians, and Mexicans. Our instructor, Alan Hershman, owned a sheep farm in Sheridan. Mr. H, as we called him, was professional, straight to the point, and the most passionate educator I ever met. He lived and slept RDAP. He was very careful not to disclose much about his personal life. All I knew about his private affairs, aside from the sheep farm, was that he and his wife were having a baby and that he visited Israel so often I thought he must have been in the Mossad.

We would challenge each other in class. He liked to use his knowledge of psychology to push our buttons and get reactions out of us, but I had a PhD in Game and was willing to put it up against any other degree. Still, the class really got to me. I thought it would all be bullshit, but it wasn't, and I started learning a great deal. One of the topics that truly touched me was the "Criminal Lifestyle" module. Mr. H wasn't buying any victim nonsense; you couldn't be on that bandwagon in his class.

At the end of the day, he always said, "Nobody pulled a gun on you to make you commit the crime; you made the choice." That stuck with me, that I was a victim until I came into knowledge of self and realized that I was the fool and fell right into the hands of my criminal mentors.

I had a lot of time to reflect, and I was forced to. One requirement for completing the program was to write a fifty-page memoir, exploring your life and the decisions you had made. We also had to write a five-page strategic plan detailing how we were going to support ourselves and survive our first six months of freedom.

One way Mr. H. got us to examine those issues was by having us study a module called "Cognizance: The Working of the Mind," which dealt with subconscious thinking, and how it helped criminals justify their crimes. It was interesting to listen to all the criminal minds in the group. Everybody had a hook to explain away why he was in prison, including me.

By the end of the nine-month program, a third of the group finally admitted that they were not victims—and one of them was me. Even if the cards were stacked differently for young African American males compared with their white counterparts, I came to understand it just meant that it was my responsibility to be stronger, smarter, and sharper than the next man.

My inmate supervisor in the bakery was a Native American named Roy-ball. He ran the baking crew. Roy-ball was a funny guy, who talked to the dough. He had this whole little rap, talking about being passionate, becoming one with the dough, how to touch it, roll it, feel it until you knew when it was ready to become the product it was meant to be.

Since I was the newbie, I had to do a lot of the tough, physical work, like carrying the bags of flour and cutting all of the bread. We made fresh bread every day, and a whole lot of it. He taught me to use both hands to roll two cloverleaf rolls at a time. It was a bit awkward at first for me, partly because I had learned some incorrect techniques along the way, but I soon caught on.

Meanwhile my dad got in touch with my aunt Eleanor, who had a lot of money. She had always been willing to pay for education for any of her nieces and nephews who wanted to make something of themselves, but she would never bail anyone out of jail. When Pops told her how well I was doing, she sent me some cookbooks.

At the suggestion of Mr. H, I started writing letters to the Culinary Institute of America in Hyde Park, New York, to get their tuition and enrollment information. I also ordered a book from the Culinary Institute of America called *The Professional Chef*. At five hundred fifty pages, it more or less covered the school's entire curriculum. I wrote away for information from Los Angeles Trade Tech. I bought *Professional Baking* by Wayne Gisslen, which became my baking bible, and *Becoming a Chef* by Andrew Dornenburg and Karen Page. I studied the books until the pages started to fall out of their bindings.

When the head baker position opened at the detention center across the street from the camp, I jumped at it right away. I was a perfectionist. Every pie, every cookie, every doughnut had to be perfect. The flavors had to be on point and the presentations had to be just like the pastries in the Jewish bakery me and my granddaddy used to clean when I was a little boy. While the other guys were just passing time, I took it all very seriously, because this would be my future.

The bakery had a big oven with rolling shelves, two proofers where the dough rose, an eighty-quart Hobart mixer, and an L-shaped steel table to work on. I considered it *my shop,* and I really put it down in there—especially on Sundays, when I made the cinnamon rolls, which quickly became my signature pastry. I tweaked the jailhouse recipes by adding extra sugar and molasses to the dough, because I liked my stuff very sweet. I would roll out a five-pound mound of dough and feed it through the sheeter until it was flat enough to roll, then I brushed it with melted unsalted butter, sprinkled on cinnamon and pecans, and rolled it tight like a log. I cut the individual rolls from it—always nice and fat. The warden and the guards loved them. The guards were always saying, "Henderson, what have you got back there today?"

I also put a new twist on my pastries using canned fruits. Through a lot of experimenting, I figured out how to use those wet fruits while keeping the dough relatively dry. The inmates used to complain when the dough was too gooey, but I got it just right. After making recipes a few times, I had them memorized and was able to alter them a bit without referencing notes. This helped me and my crew to be more productive, and to have the time we needed to make a really quality product—at least for a prison.

Some kitchen foremen let us take extras back to camp. They'd search us going in and out of the kitchen, but overlooked whatever we could carry in our hands. It was common practice in all prisons; the guards knew there had to be some extra payoff for the hard work we put in in the kitchen.

I began working six days a week, sometimes seven. Time went fast in the bakery. I was closing in on the last three months in RDAP. I had just ninety days and a wake-up left.

No one visited me while I was in Oregon, but by that time, I was getting so much mail they started calling me The Mailman. I heard from many old friends, and former girlfriends started communicating with me for the first time in years because they heard I was close to getting out. It was too late for most of them who had forgotten me early in my sentence, and I wasn't going back to the life I had.

I stayed in touch with Friendly through a friend of his named Delores who I met when she came to Nellis to visit him. Dolores loved me like a little brother and said she kept pictures of me around her apartment. She was also a close friend to one of Friendly's nieces, Stacy Womack—daughter of the late great Mary Wells and Cecil Womack (who is married to Linda Cook, the eldest daughter of legendary soul singer Sam Cook). Stacy had seen my photo at Delores's house one day.

"Damn, he's fine!" she said to Delores (or so I was told).

"That's my friend Jeff," Delores told her. "You know, he asked me if I knew any nice women to introduce him to. He's a good man, he just wants a fresh start. He's looking for someone to help him get acquainted with the outside world again."

"Well," Stacy said, "he *is* good looking. I'll write him a letter."

Over the course of several letters, I gave Stacy the whole jailhouse spiel: I was reformed; I had a game plan for when I got out of prison. On the phone, I used my old gift of gab. Hell, after almost a decade being locked up you can charm almost any woman on the telephone just out of sheer determination.

Stacy sent me a couple of pictures of herself. She was beautiful, as well as sharp and articulate, and she had gone to college, which was a big deal for me. I wanted a woman who did her own thing, made her own way, a woman who respected herself too much to live off some man, especially not some knucklehead thug of a man. No more baller girls for me.

Our relationship became more and more serious as my release date approached. On the one hand, I wasn't sure that I wanted the challenges that come with a committed relationship when I'd just served almost ten years in prison. At the same time, all my life, I had always needed a woman beside me. I had always felt I needed someone to love me. And I was afraid of being alone on the outside.

Stacy told me that Friendly had advised her to be cautious with me. He was my boy and all, but he knew that I used to run a lot of women back in my days on the streets. He didn't want his niece to be one of many women if I went back to my old ways. By then Stacy and I had already begun to fall in love with each other, even though we had never met. I asked Delores to tell Friendly that I would respect Stacy and treat her right, and I meant it.

Closing in on graduation at RDAP, everyone was excited. Even though we wanted out of prison, we feared the unknown. I had some loot put away with my father and didn't have immediate money worries, but I didn't know if I could make it out of the prison system and on the streets not doing the only thing I had ever been good at, selling drugs.

Mr. H called me into his office one Thursday after taking a two-week leave while his wife had their baby. During that meeting, he showed me a *USA Today* article about the top African American chefs in the country. There was Patrick Clark, the executive chef of Tavern on the Green; Marcus Samuelsson, who had his own restaurant in New York; and, Robert Gadsby, who was about to open a restaurant in Los Angeles. Mr. H thought I should get in contact with these chefs for mentoring and a possible job once I got released. Clark and Samuelsson were in New York City, so they were out because I couldn't leave California while I was on probation. Gadsby and his soon-to-open sixty-six-seat namesake restaurant in L.A. was a possible opportunity.

The *USA Today* article fueled my passion to reach a higher level of success in the food world. I became obsessed with culinary knowledge and the reimaging of Jeff Hard Head. I immediately penned a letter to Chef Gadsby. I wrote him that I was currently incarcerated, but that I had found my love for cooking while in prison and that, upon my release, I would like an opportunity to work for him. I explained to him that I was a hard worker and that I had redeemed myself for my past criminal behavior. I told him that, more than anything, I wanted to become a chef.

I also wanted him to know how proud he, Patrick Clark,

and Marcus Samuelsson made me feel. I never realized, I wrote Gadsby, that brothers actually cooked cuisine on that level, because in all the food magazines and books that I read during the last few years I had never seen any African Americans in them.

"But when I saw you guys," I wrote, "in *USA Today* of all newspapers, I began to dream bigger."

I knew then, by just writing that letter, that I would be more than just a cook at some fast-food joint. I sent the letter off and waited patiently every day for some reply. It never came. I wondered if he ever even received my letter and, if so, whether he read it.

TWELVE

THE GREAT GADSBY

I spent the day before my release socializing with some of the guys I'd become friends with at Sheridan and passing out all of my personal belongings to some of the brothers. After finishing up my last workout, I showered and started walking the track to mentally prepare myself for the big day. The next morning, I went to chow at 6:00 A.M. so I didn't see most of the fellas. I didn't want to rub it in their faces that I was going home while they still had time to do. That's traditional for most inmates. You don't say your good-byes; you just disappear off the yard one day.

After breakfast I was issued the release clothing that Stacy had sent, a one-way ticket on Alaska Airlines, eighty bucks, and my itinerary with directions to the halfway house in L.A. By nine in the morning, I was en route to the airport.

Stacy had wanted to come up to Oregon so we could fly home together, but I told her not to in case the Feds were watching. I was out of prison, but they still had papers on me, and my instructions were to go directly to the halfway house.

As soon I got to the airport, I headed straight to the candy store—I'd been craving jellybeans for many years. I was at the counter buying a half pound of them when someone tapped me on my shoulder. I turned around and there was Stacy.

It was the first time I'd ever laid eyes on her in person. She was fine as hell.

"Damn! What're you doing here, baby?"

We sized each other up, looked each other up and down. We hugged and kissed.

"Baby," she said, "I got us a hotel room. It's right down the street. We've got an hour and a half until the plane leaves."

"Oh, I can't leave, Stacy. Someone could be watching me right now."

"It's okay, baby. Quit being paranoid. Come on!"

What could I do? I gave her a big grin. Of course she didn't have to do any more convincing. We raced out of the airport and into a cab. In the room, Stacy surprised me with a single red rose. It was the first time in many years that I'd been with a woman.

After that quick rendezvous, we made our way back to the airport and Stacy held my hand tightly as we waited for our flight.

When we landed at LAX, my father, my sister, my son Jamar, and his uncle were all there. While it was a great feeling to be back, I was very tense about my newfound freedom. My biggest fear was a sense of the unknown. I couldn't shake the feeling of impending failure. I would have to have self-talk, regroup, and get strong again.

My family drove me to the halfway house, dropping me down the block because I was supposed to have taken a cab. So right from the start I was on edge because I was already breaking some of the rules of my early release.

The halfway house was an old apartment building with a fence around it adjacent to a Taco Bell. It was on the ho stroll right on

Century Avenue and Prairie across from Hollywood Park, in a neighborhood that was nothing but hookers and dope fiends.

When I checked in, I was issued a key to my four-man room and met with a caseworker who went over my "Employment Plan of Action." The first thing I asked her was if I'd be given a furlough to leave the complex and look for work. She told me it would take a couple of days but that I'd eventually get an eight-hour pass to start applying for jobs. I was in a hurry to make my way to Gadsby's restaurant. Even though he never answered my letter, my mind was set on it.

As soon as my furlough came through, I made my way to Gadsby's on the city bus. When I got off the bus, I stood out front with the *USA Today* article about Robert Gadsby and a copy of the letter I'd written him, getting my head together.

It was lunchtime inside and not too busy. I noticed a three-man crew of Latinos working the line. Chef Gadsby was behind the sushi bar being photographed. Robert was probably the first black chef in America to own a sushi bar. Hell, I didn't know a thing about sushi. I'd read about it in prison but had never tasted a single piece of it.

When the host approached me, I said I was there about a job and waited patiently for Robert to meet with me. It was forty-five minutes before he came out of the kitchen. I didn't mind the wait, as I was too excited and enjoyed just standing around watching.

Robert was about five feet ten inches, bald, with a jet-black complexion, a neatly trimmed mustache, circular black-rimmed glasses, and it looked like he used about half a jar of Vaseline on his head. The man was in tip-top shape and very polished. He'd later tell me, "Never let yourself get fat and greasy like a lot of the chefs you see today."

One thing that struck me about Chef Gadsby was that he was wearing clogs. I always thought clogs were for women. I liked his style, but I could never imagine wearing clogs. People would think I got turned out in prison.

Despite the shoes, Robert had a presence of a four-star general. He was suave and very articulate. The article said he trained with some of the world's top chefs. I was intimidated, but I kept it off my face.

"Yes? What can I do for you?" he asked in a heavy British accent.

"Mr. Gadsby," I said, "my name is Jeffrey Henderson. I wrote you a letter about three months ago, asking for an employment opportunity once I was released from prison. Before I tell you anything else about myself, I want you to know that I'm a nonviolent, first-time offender, and I fell in love with cooking, discovered a passion for it, while I was incarcerated."

I then showed Robert the *USA Today* article, telling him, "I was so inspired and moved by your story. I knew I wanted to work for you as soon as I read it."

"Well, I don't have any positions open right now," he said. "I just opened the restaurant and we're presently fully staffed."

"Do you know if you're going to have anything opening up anytime soon?"

"Well, you never know in this business. Some people stick around for years and some people just don't make it. You have to keep checking back with me."

I said, "Okay. Thank you much for your time," and left the restaurant. I was a little disappointed; I guess I felt that just because I was black and looking for a break that he would have to offer me something, anything.

None of the other restaurants I visited had any openings for me, either. So I went with what a temp agency got me: part-time work as a trashman in the city of Glendale. That job was tough, backbreaking, and I was paid minimum wage to hang off the back of a truck picking up rich folks' garbage. When you collect trash for the rich, every house has nine or ten trashcans on the curb. I became more determined than ever to find work in a restaurant.

I kept after Gadsby, visiting him at least twice a week for a month and a half before he finally decided to take me seriously. One day he asked me to take a ride with him to a community center over in Watts, where he had started to do some work with inner-city kids. Robert conducted a cooking demo for some of the kids and lectured them about locally grown products.

After we toured the center, he asked, "So when can you start working?"

I hadn't had such a wide smile on my face in a long time. I told him, "I can start right away, sir."

"All right. I want you to start on Monday."

"Yes, sir. At what position?"

Robert told me I'd be starting on dishes; I thanked him and promised that I wouldn't let him down. Then he confided something to me that I would remember for the rest of my career.

"I was initially hesitant to offer you any position at all," he said. "To be honest, I've had some bad experiences with black men working in my kitchens."

I assured him, "I'm not going to come in with some slick bullshit, Chef. I'm a hard worker, I'm teachable, and I want to learn."

Back at the halfway house, I immediately got on the telephone and started calling my family, especially my dad, to let them know I had just been hired by one of the top chefs in America.

My relationship with my father was strained, though. When I'd first gotten out, I had asked him about all the cash and jewelry I'd left with him almost ten years earlier. He was hesitant to talk about it and kept putting me off, saying, "We'll discuss it when you come over and I'll give you your things, but there's something I have to tell you."

As soon as he said that, I knew it was all gone. My father told me that having all that cash around had become like an addiction. He had been borrowing from my stash for years, telling himself he would be able to pay it back before I got out. But he kept dipping into it until there was only $25,000 left and some old-school jewelry that wasn't worth much. If I'd known he was going to piss off all my money, I'd have left it with my mom. Hell, she was working two jobs and struggling more than anyone while Pops had his own business, a home, and a wife who worked. I knew my father loved me, but then there was a part of me that wondered whether he visited me in prison so often because he loved me or if he just felt guilty because he was living off me while I was locked up.

At the same time, the bond between Stacy and me kept growing stronger every day. She was at the halfway house every night, making sure I was taken care of. She even made sure I had my first real dinner at the halfway house. She brought me lasagna with an avocado and tomato salad and Italian dressing that she and her cousin had prepared. She even brought sparkling apple cider and champagne glasses to drink it in.

T-Row showed up a couple months after I got out and offered to take me shopping. By then I was getting weekend passes for recreation, and we had a lot to catch up on. T briefed me on what was going down on the streets, but I really didn't care. I appeared interested, though, because I didn't want T to know how much I had squared up. He took me to Berini in the Fox Hills Mall and bought me about $1,000 worth of clothing, and then he put some cash in my pocket—about $500 bucks. T had aged a bit, and he seemed to have slowed down his game somewhat. I guess he got tired of doing time. After that, I didn't see much of T. I wanted to stay away from everyone from my past because I wanted to be sure I was strong enough to resist my old criminal urges—and I always felt like the Feds were watching.

One by one my old homies all started to get in touch with me; eventually even Carmen called. I didn't know what to say to her, especially about how she had helped me out, so I didn't say much. We had been really close once and had gone through a lot together, so I hoped that we'd have the chance to be friends again one day. But I thought that she was probably still connected to the game and so she was as off-limits as any of the homies I used to run with. I'd have to keep my distance.

The only person from back in the day who didn't try to get in touch with me was Dana. She'd moved on and was married with two kids. I attempted to contact her once and she got word to my father that if I tried it again she would tell my probation officer I was harassing her. I didn't need that kind of trouble.

I knew that I'd soon be able to leave the halfway house and be granted home arrest if I stayed on good behavior for the rest of the first three months. The original plan was for me to move in with my grandparents, but Stacy started putting major pressure on me to stay with her. She told me I needed love and affection day in and day out. That sounded good, but I wasn't 100 percent sure that it was what I really wanted. I'd gone to prison as a youngster who'd never had a monogamous adult relationship. I hadn't had so much as a single day of freedom before I'd suddenly found myself in one. In many ways, my heart had hardened through the years, and I just wanted to get my culinary career on the road. So I left the question up in the air for the time being and concentrated on the work ahead of me.

The night before I was supposed to start my job at Gadsby's, I tried to remember all the recipes I had learned in prison, not knowing that, aside from Friendly's Fried Chicken, I would never use any of them again.

Stacy picked me up on her way to work at the West Coast sales office of *Nation's Restaurant News,* where she was an assistant to one of the sales execs, at seven in the morning, and dropped me at Gadsby's. I was three hours early because I wanted to case the place, vibe it out a little before the rush started. I peeked through the window to psych myself up emotionally and mentally.

I was out there for about an hour when Robert pulled up in a white pickup with a bed full of groceries and restaurant supplies.

He said, "How are you doing today, Jeff?"

"Fine, sir. I'm ready to go."

I showed up in jeans and a button-down shirt, so Robert took me in the back and gave me a white chef jacket, black pin-striped cook's pants, and an apron. While in the bathroom changing into my first chef gear, looking into the full-length mirror, I felt proud to look professional in that double-breasted crisp white chef coat. I hadn't even touched my first sauté pan, or reduced a sauce yet, and I felt like I had arrived. I was pumped. I immediately went out and unloaded everything from Robert's truck to show him that I was eager to get to work. I knew I had to work faster, harder, and smarter than the Latino guys he had working for him. They were my competition.

After we unloaded the truck, Robert gave me the grand tour. He showed me my work area and told me what my daily responsibilities would be: keep the restaurant swept and clean, as well as the bathrooms, toilets, and sinks. Of course I would shine at those tasks, having been trained in the Tidy Bowl arts since childhood.

Robert then showed me the dishwashing machine, which was new to me, so it would be the first thing I'd have to learn, and I'd have to learn it quickly. I also had to train my hands to be gentle with wine goblets and champagne flutes because they were so delicate. In prison, everything was plastic, so I wasn't used to dealing with glass. Hell, I'd just gone nine years without once eating with a real fork.

Robert asked me if there was anything else I needed to know, I said, "No, sir," and it was on.

I introduced myself to the Mexican crew as they started showing up. They were cold toward me and, I thought, resentful. Maybe they looked at me and figured I was somehow connected to Robert or a member of his family, but I was making minimum wage just like any of them. I guess the difference was that I would have done it for free. All I wanted was to learn.

Over the next few weeks, every time I got a break from the dishes I'd go over to the line and observe Gadsby and the other cooks at work—sautéing, braising, and plating up elegant towers of fine cuisine. He was an artist and I admired the way he moved in the kitchen. But one thing I didn't understand was his aggression, the way he spoke to his crew. He was very hard on the Latino guys, barking at them whenever he wanted them to do anything. Eventually, he started giving me the same treatment and it got to the point where I thought I might lose my temper and beat him down. Then he told me something that I will never forget.

He said, "When you work for me, I'm going to train you to take people's jobs, and hopefully you'll never lose your job to the next guy down. Every day that you work in this business, someone who wants your job will always be lurking over your shoulder. So you have to bring your A-game to the kitchen every single day—whether it's cleaning chickens, deboning fish, creating that night's special, or just scrubbing toilets. Don't take it personally. I am going to make a chef out of you." Then he patted me on the back.

He also promised that there was nothing I couldn't learn under his tutelage as long as I stayed focused. "No one works one position in a freestanding restaurant," he said. "If you work here, you have to work sauté, hot appetizers, the broiler, prep, and dessert. You have to learn everything. Freestanding restaurants, as opposed to the big hotel restaurants that are run with big money, don't offer the luxury of working just one position. Here you wear multiple hats."

Finally, he told me, "Never forget these three things: speed, taste, and presentation. Once you master those essentials, there is nobody who can stop you."

I took his advice to heart and did my best to not let his tough attitude get the better of me. Before long, he moved me up to the prep station. Banging alongside the Latinos, I studied and watched their every move. They knew that I was going to work as hard as them and they didn't like that. But my aggressiveness and intimidating presence kept them in check. I was grinding for a paycheck, just like them, but I was also focusing on honing my skills. Besides, Robert had told them to train me.

Almost as soon as I started my next position, Robert had some more advice for me.

"Don't try to mimic me; become your own chef," he said. "It takes time. Be patient. Your time will come, but it takes many years to develop the supreme touch."

I said, "Chef Robert, what do you mean 'the supreme touch'?"

"It's the way you handle the food. Jeff, you were in prison all those years lifting those big ol' dumbbells and such. But don't you notice that every time you put the garnish on top of the fish, it falls off? You need to relax—you're manhandling the food. You have to handle food with love and respect, and then the food will respond to you. You'll get it. Don't worry."

A few months later, my touch became polished enough that

Robert gave me an opportunity to help with the desserts and pastry, my favorite place to work in any kitchen. One of the things I told him from the beginning was that I wanted to be a pastry chef.

"You can always find a cook," I said, "but you can never find a good pastry man or a baker. And in all my reading, I've never heard of any black pastry chefs, so I figure my best shot at making it in this business is to go into an area where most brothers don't go."

Robert quickly defused that line of thinking. He said, "Jeff, in this business, it's not about the brothers. It's not about the Latinos or the whites. It's all about talent, it's all about passion. Get out of the brother business." Then he added, "And forget about soul food. No one cares about soul food. It's simply one of the most disrespected cuisines in the world. You have to learn progressive American cuisine. That's what I do, and that's what you will learn from me."

I agreed on the one hand, but on the other, it was me in the trenches, getting no respect from the Latinos. The cold stare downs, no teamwork, and having the feeling of being in a foreign country made me really question whether I could lose my identity entirely. I was warned prior to my release that the Latinos had the L.A. kitchens on lockdown and that I would have to bang hard to get respect and get a job in the top kitchens. As in my former life on the streets, I knew I had to take control and use my leadership skills to find a way to convince the Latino cooks that I was just like one of them and our struggle to get ahead was the same. I used my PhD in Game to convince them to let me in their crew.

When Robert moved me to the dessert station, his baby, I knew he was proud of me in his own way. While Robert had no connection to the hood and he'd never admit this, I knew that he wanted to help me because I was black. He was giving me opportunities to move up that few others got in his kitchen and he gave me private cooking lessons when the rest of the crew wasn't around. He made it his mission to share what he learned from Joël Robuchon, Alain Ducasse, and Thomas Keller. In turn, I wanted every piece of advice I could get out of his bald head. I wanted him to be proud of me. I worshiped every dish he made and savored every flavor he introduced me to.

Robert's food was progressive American, very simple and clean with bold flavors. The portions were small but with intricate presentations. They had vibrant colors and the ingredients were arranged as though Rob was setting a stage. Robert was very particular how everything looked, but he was careful not to make anything too complicated. More than anything, though, he reminded me that it was all about creating great-tasting food and to pull that off we had to make sure that we timed the cooking of each course perfectly.

I'd been working at Gadsby's for almost year, had left the halfway house, and was living with Stacy in her condo in the upper-middle-class black neighborhood of Baldwin Hills. She had finally convinced me to move in with her. I enjoyed living with her, but I had problems with her brothers, who were constantly hanging around the apartment. Her brother Meech was a record producer and her brother Shorty was an up-and-coming rapper. Meech was laid-back and never talked much, but Shorty was as young and hardheaded as I was before I caught my case. I told Stacy I was concerned about Shorty coming around because I was still on probation and could catch another five years just for being around weed—and I didn't want to be around all his homies from the hood and the trouble they could have gotten me in.

I didn't want to come between Stacy and her brothers, but I finally had to tell her to make a choice: Shorty or me.

It all came to a head when Shorty and I had a confrontation about a cordless telephone I bought for the house that almost ended with us having a fistfight in the living room.

I told him, "Shorty, I'm not trying to have any beef with you, man, but you know what? I get up and go to work every day like a man. I make a contribution to this place. Every time you're here, you leave the lights on all over the house, you leave the place a mess. And you're always on my phone. You have to bring something to this house. I don't care if it's eggs or a loaf of bread. You need to make a contribution."

Shorty was like, "Fuck that, nigga! This is my sister's house. I was here before you was here!"

"Yeah, but your sister's my woman. As a matter of fact," I said, "Stacy, come out here."

When Stacy came into the living room, I put it to her simply. "You're going to have to make a choice right now on who's going to stay here in this house. I'm telling you right now: I cannot stay here with this type of bullshit going on."

"I'm sorry, Shorty," she said. "I love you, but I have to have peace in my home. If you keep smoking weed and bringing all them guys over here, you're gonna have to go."

Then he lashed out at his sister. "You're gonna defend him the fuck over *me*? I'm your blood *brother*!"

"It's not that I'm defending him over you. But this is my place and you have to respect that. This is my man and he's not going anywhere. And I need to be happy, too."

Shorty made his decision. He left.

After that, things were a lot cooler between Stacy and me. We had the whole domestic living thing down, but I never felt entirely comfortable with it. A part of me still felt institutionalized. I was very paranoid and security conscious. Mostly, though, I had become selfish.

In prison, everything is about you—your little cell, your few belongings, your survival. In prison, there really was no one else in my world. I thought only about how to do my time and survive. Even when I was in bed with Stacy at night, I curled up in a little section of the bed with my back against the wall like I was still in the top bunk. If she put her hands on me I flinched because I was so used to staying on the defensive even in my sleep.

Showing her affection began to be a problem as well. I knew I would have to take time to loosen up. I've never really been a lovey-dovey type of guy, and being isolated in prison for so long had made it even harder for me to show affection or express my deep feelings to a woman. It was tough for me, and instead of facing it, I focused on my mission to become a chef.

My edginess bothered Stacy, but she was happy to have a man who was responsible and was following up every promise I had ever made to her in my jailhouse love letters: to be down with her, never to leave or mess around on her, and to become successful as a chef.

She always wants me to be more affectionate, but, the truth is, I don't know if I ever will be. My mother and father were never that way with me. I knew they loved me, but they loved me the only way they knew how. Hell, their parents were the same. To this day, I work as hard as I can to break that cycle. I always tell my wife and children that I love them, and I show my children extra affection by kissing and holding them every day.

The restaurant started to really pick up. We banged out sixty to seventy covers a night during the week, a hundred and twenty on the weekends. The dessert station always got slammed toward the end of service, and the pressure was heavy. The supreme touch was very critical for desserts; positioning edible garnishes and stacking Robert's towering confectionary creations was a challenging and precise task that could give the most even-keeled cook a panic attack.

Every dessert was handmade with the freshest locally grown ingredients. The first dessert Robert taught me was his Granny Smith apple strudel. I had to peel and dice apples into tiny squares and then quickly sauté them with a little bit of white sugar, ginger, and a hint of fresh cracked white pepper. They were removed from the heat and left to cool on a half-sheet pan. Next, I rolled the cooked apples in phyllo dough, then brushed the dough with clarified butter and sealed it. I'd sprinkle it with graham cracker crumbs and bake it in the oven until the dough became a beautiful flakey gold. The aroma was just like apple cobbler. The pieces were cut and stacked on a bias and served with some of Robert's homemade French vanilla bean ice cream. Robert always took his dishes over the top, and this one was drizzled with rich, warm butterscotch and garnished with mint and a touch of powdered sugar.

His desserts were so good that I couldn't wait to eat the scraps after I plated them up. I had to be careful, though. If Robert caught any of his staff eating his food, he gave them a verbal ass whipping or fired them. He always said when we tasted a product it was for flavor only. Robert made sure he made every dime in profit off his inventory. All the food the staff ate was bought with separate money. I never even saw Robert eating his own food except to taste it for quality. He also didn't believe in eating before service; to Robert, that just made you bloated and lazy. I quickly adapted that M.O. At the end of the night, he would order Jamaican food from Juicy's or get chicken breasts from Ralph's down the street and we would make some pasta to go with it.

One day I brought a piece of my own chicken from home, sea-soned it up with Friendly's spices and put it in the deep fryer at the end of the night. I didn't see anything wrong with that, but Robert got right on my ass.

"Don't ever put fried chicken in my deep fryer," he told me.

I was, like, "Chef, I brought it from home. This is nothing from the restaurant."

"I don't care! I do not want fried chicken in my restaurant. Worse, it changes the flavor of the oil, and that affects the flavor of everything else we prepare in it."

"All right, Chef. I'm sorry."

"And how many times do I have to tell you that you have to break yourself away from that soul food anyway, Jeff? You need to start broadening your palate if you want to have any hope of becoming a real chef. You need to start eating and tasting new things. That is the only way you will ever grow in the food world."

Robert was often hard on me, but he wanted to make sure he taught me all the things I needed to know to make it in the world of high-end cuisine, and I knew that I needed a lot of help. He was so hands-on that he would come off the line and wait three or four tables a night. At Gadsby's, he was the owner, chef, server, and sometimes even the host. He took the tips he earned waiting tables and spent it on sampling the food at restaurants all over Los Angeles, and he often took me with him.

I must have tasted every type of cuisine there is on those trips, including French, Italian, Chinese, Korean, Thai, and In-dian. I was afraid that, sooner or later, I was going to have to try sushi, and that was just something I wasn't ready for. Robert loved all things Japanese. Many of the dishes on his menu had a Japanese aesthetic, but that wasn't the same as actually eating raw fish. There was a popular eatery on Third Avenue called Sushi Roco and Robert insisted that I check it out with him. I was thrown the second we walked in and all the sushi chefs yelled, *"Trashaimase!"*

Robert ordered several types of sushi. I didn't want to embarrass him or the chef, so when the meal came I put a piece in my mouth and pretended to enjoy it. Robert had also taught me a useful trick. He never wanted to send anything back, so he'd stealthily spit anything he didn't like into a napkin and put it under the table. As soon as Robert looked away, that's just what I did.

With Robert, everything had to be perfect. Every dish, every piece of fish seared properly, every vegetable blanched and seasoned perfectly, every chicken breast roasted with herbs as exactly as he showed us. Nothing got to the tables unless it was just right.

Even Robert's bathrooms were perfect. He dusted the plants and sprayed olive oil on them so they would shine. His walk-ins were immaculate. You could eat off the floors. Robert's operation was so tight that when the health department came to inspect, they'd look through the kitchen door in awe and leave.

I was mesmerized, inspired, and motivated by him. He not only taught me how to cook, but how to behave, how to conduct myself among the top professionals in the industry. He talked to me about my walk. "Your walk is too hard," he said. "You need to ease up a bit—people take notice."

Nothing was cool about Robert. You'd think he was a white guy in blackface. A lot of professional African Americans in L.A. didn't understand him. It may have been his British background. But I thought I understood him. He just wanted to be recognized among his peers, and his peers were white professional chefs. My take on the whole culinary game was to earn respect among my peers as well, but my peers also included people in my community, and my family. My mission was to open my own restaurant and provide opportunities for people from the community.

One night Robert took me to Wolfgang Puck's restaurant, Spago, in Beverly Hills for a seven-course tasting prepared by Lee Hefner, the chef de cuisine. He conducted traffic from a high stool in the open window of the kitchen like some kind of field general.

After most of the dinner crowd had gone, Wolfgang and his wife, Barbara, came out and sat with us. I was proud to meet him, but my feelings soured when Robert said he wanted to take part in a Meals on Wheels event Wolfgang was putting together. It almost seemed like Robert was trying too hard to be accepted by the uppermost echelon. Wolfgang gave him the brush-off.

"Talk to my people," he said. "We'll discuss it further." It was just *blah-blah-blah*. There was no high-end black representation at the event.

When we left the restaurant, I began to play the race card. "Doesn't it bother you that there are no brothers in any of these kitchens?" I asked him. "Not even the dishwashers."

Robert rejected that argument once again. He didn't believe there was any racism.

As if to prove a point, Robert kept going out of his way to expose me to the power players in Hollywood and Beverly Hills. Working alongside Robert, I helped prepare six-course tasting dinners for people like Laurence Fishburne, Madonna, the Jackson family, and Stevie Wonder.

I never talked much around these people, just tried to listen and learn. As polished as I was becoming, I was still scared I would say the wrong thing, mispronounce a word, or use it in the wrong context. Still, I was very proud to see a black man conducting himself the way Robert did among influential people. Before Robert, Kevin X and the brotherhood in prison had been the most intelligent and most articulate black men I'd ever met, but now I knew that I had to develop my conversational skills until they were on par with Robert's.

I worked on the way I walked, cut back on how often I lifted weights in order to slim down and make my appearance less intimidating, and learned to smile more—all to defuse my prison demeanor. I even bought a pair of clogs and a Brigard chef's jacket. I was probably the only low-level cook in L.A. wearing a $150 Egyptian cotton chef's coat. Robert used to tease me about it all the time: "If nothing else," he'd say, "you'll win best-dressed chef." I may have been a beginner, but I looked like I had been a chef for twenty restaurants.

It was October 2,1997, and we were celebrating my first year out of prison anniversary at Bertha, one of the best soul food joints in L.A., when Stacy began telling me that she wanted us to have a baby.

I had put that baby talk off as much as I could, but she wasn't about to let it go. Over the next couple of months she kept bringing it up, telling me that I needed to have another child, until I finally agreed.

Lordy B! She was knocked up before a month passed. I was, like, "Holy shit! Am I ready for another child? Well, it's not like I have any choice."

I felt guilty, because I had failed my son Jamar. He was just five when I went down. Before that, all he ever knew was that his dad was rich as hell and drove flashy cars loaded with loud music. I knew deep inside I owed him time, love, and fatherhood, but I just didn't know how to give them to him. It wasn't as if I had an example of good fathering my whole life. I owed an awful lot to his stepfather and his mother for keeping him strong and away from my past. Having more children would give me a chance to redeem myself for my failure with my firstborn son.

The more I thought about it, the more excited I got. And, Stacy? She was happy as a jaybird. A couple months later, we saw in a sonogram that we had a baby boy on the way. I was happy, not only because I was having my second son, but because this time I would have the opportunity to raise him from newborn to manhood. I called Jamar and told him about his new little brother on the way. He responded like I thought he would; he was cool, but not cool with the news. My joy was dampened with shame.

On the job front, I was learning so fast at Robert's restaurant that it was like an overload of information. Every extra dollar I had, after I gave Stacy my half of the expenses, I used to buy new knives, china, and baking supplies. I added a new knife every payday until my collection was complete. I saved up and got myself a two-quart KitchenAid standing mixer.

Every chance I got, I would experiment, perfecting Robert's dishes and then trying to develop my own. Everyone who came to our home became my guinea pig. I couldn't wait to show Stacy and our families my revised versions of Robert's desserts and my new creations. Stacy was blown away by the drizzled chocolate, even though I used Hershey's syrup instead of the expensive chocolate Robert used at the restaurant. It made me happy to be praised for doing something I was good at, especially from my family.

When I was in prison, Mr. Hershman had worried that I'd struggle in the outside world because I wouldn't get the pats on the back that I was used to receiving in detention. Learning to self-praise was helpful in keeping up my esteem whenever it began to slip, yet I always needed to be validated, to prove to my family that I had changed. In many ways, I still needed to prove it to myself. I knew I had paid the price for my criminal deeds, but the future was still a question mark.

Four months into her pregnancy, Stacy decided that we should get married. I loved her, but it felt like she was piling one responsibility after another on my shoulders while I was struggling every day to build a stable, successful career. Sometimes I wondered if she really understood the pressure I was under. We were driving on the freeway when Stacy brought up marriage again. This time, however, she brought up the teachings of Kevin X and Imze: respect your woman, your brethren, and your family.

She hit my weak spot. I pulled off the freeway and we immediately headed to the county courthouse and got our marriage license. We got in line, went before the registrar, and I was rubbing Stacy's belly as I said, "I do."

Not long after our son Jeffrey Jr. was born, it quickly became clear that the minimum wage Robert was still paying me just wasn't going to cut it any longer, but I felt guilty about asking him for a raise. To me, the skills he had taught me were worth much more than I was earning. At the same time, here I was a father of two and a husband and I didn't even have health insurance.

By that time, I understood the culinary game well enough to know that the only way someone with my limited professional experience could earn a decent wage and get health insurance was to take a position at one of the big hotel chains.

Robert once told me hotel chefs get no respect. "They're

big fat men with handkerchiefs around their necks, wearing tall toques, and doing mass-production banquets." The only chefs who got recognition, he said, were the ones in freestanding restaurants who operated independently of the massive food industry. Stacy wasn't buying Robert's story. She had bigger plans for us.

I was worried about coming up with a way to tell Robert that I was moving on, but, as it happened, I didn't need to.

One Saturday night his standing mixer went missing and he went on a rampage. Robert is a man you don't steal from. He would know if so much as one piece of salmon or a slice of foie gras went missing. He searched all over. The guy even went out back and checked in the Dumpster. After ranting and raving at all the Latinos, he approached me.

"Jeff," he said, "my mixer's missing."

"Robert, I wouldn't take your mixer. I have my own mixer at home. I would never take anything from you. You're like a big brother to me. I have your back. I watch the Latinos for you and make sure they don't jack you."

"But it was here earlier and the Latinos take the bus home— you have a vehicle."

It was midnight, but I whipped out my cell phone and called Stacy and said, "Could you get my mixer and bring it down to this fool's restaurant with the receipt?" When I got off with Stacy, I told Robert, "I was never a nickel-and-dime hustler. If I was gonna get you for something, I'd get you for money."

Even after Stacy came by with my mixer and receipt, Robert continued to insinuate that I'd stolen from him. I didn't argue it any further, just told him I'd be finding another job. Despite the strain between us, Robert gave me his blessing and wished me the best.

THIRTEEN

~~~~

WORKING THE PASS

After leaving Gadsby's, I took a position as head baker at Patterson's Coffee House in Culver City to make ends meet while I peddled my resume to every hotel on Century Boulevard, near LAX airport. The Hilton, the Westin, and the Sheraton all shot me down. The Latinos had those kitchens all sewn up, and there were no welcoming arms for a convicted crack dealer from South Central who had just finished nearly a decade in prison. The last place I hit was the Marriott.

Sitting in the lobby, figuring out my approach, I noticed this big brother about six feet three inches with a tall white toque on his head and white kerchief around his neck, with white pants and white shoes. He was what I pictured when Robert had spoken about the cliché hotel chefs. I immediately jumped up and approached him.

"How are you doing today, Chef? My name is Jeff Henderson."

"Chef Sterling Burpee," he said in a very professional way.

"Sir, I've been trying to get a job on this strip for a couple months now and I can barely get an interview."

"Did you fill out an application?" he asked.

"Yes, sir."

"Okay, well, I'll see if I can pull it and get you an interview."

"Thank you," I said. "I'd really appreciate that, sir. But there's one thing I need to explain to you."

"And what is that?"

"I'm a convicted felon. I spent some time in prison for drug trafficking, but I'm not a user, I'm nonviolent, and I'm not a thief. I worked with Robert Gadsby and I will guarantee you that you'll be pleased with my work."

We spoke about Robert a bit, and about what I did while working at Gadsby's. Robert and I had patched things up after I left, so I was sure he'd give me a good recommendation.

The chef, Sterling, told me to come back in two days. He set me up for an interview with his head chef, an old German man. I talked about my commitment to the industry, my love of and passion for cooking. He asked me a few questions and seemed to approve of the answers.

Once I passed the drug test, they offered me the baker's assistant position. I was happy, Stacy was happy. I felt I was bringing some financial stability to my family.

The bakery crew moved fast as hell. They had a very efficient system in place and it took me a couple of weeks to get into their flow. I had to deal with communication difficulties, as well, because these guys didn't speak English—only Spanish. On top of that, there was added tension because, as usual, they wanted their cousins and friends to get any job that opened up. Besides Sterling, the only other black man in the kitchen was this Jamaican sous-chef who never did anything but run his mouth about his big plans.

Sterling was a consummate professional, and a true-blue, diehard Marriott man. He knew all along that I'd have some problems breaking in. It was his way of testing me to see if I'd lose my top. They all knew I was an ex-con, so I knew it was critical that I behave myself as professionally as Sterling.

It took just a couple of months to prove myself in the bakery. After that, Sterling moved me to banquets, working under this chef named Patty, who had a degree from one of the fancy culinary schools. I never actually saw her cook. She seemed to spend all her time in the office on the computer. In the banquet kitchen, I was able to use some of the skills and recipes I'd learned from Robert and from a few books. I was an instant star. I was recognized by the food and beverage director for my work and was told great things were in the future for me.

After three months, upper management asked if I wanted to join a task force that was going to open a new Marriott Hotel on Coronado Island in San Diego. I was honored. Marriott had recently purchased the four-star La Meridian hotel on an island overlooking Coronado Bay. Our job was to flip all the standard operating procedures to Marriott specs over the next ninety days. There were two hundred fifty guest rooms and two restaurants: L'Escale served breakfast, lunch, and dinner; Marrious was an award-winning fine dining restaurant.

The Marriott opportunity was one I couldn't afford to blow. They were only paying me $8.50 an hour, but I knew this San Diego assignment would allow me to jump up several levels quickly.

I went down ahead of my family, checked into the hotel, and met my new boss, Sarah Bowman, a middle-aged white woman who was wearing a toque, handkerchief, and a ponytail. She had a tough aura about her and I was a little nervous about being part of this task force, but I was ready to be a soldier for Sarah and do whatever was needed.

Sarah was a great cook, a great motivator, and something of a visionary. I immediately started learning from her. Her chef de cuisine was a Chicano named Bobby from Arizona. He was going to run the fine dining kitchen and I'd run breakfast, lunch, and dinner service at L'Escale.

Any sense of intimidation I'd felt went out the window when I saw the menu. I couldn't wait to instill it with some of what Robert had taught me, maybe even borrowing a dish or two of his. Little did I know, Chef Sarah already had her own vision for the restaurant. Within the first three days, I learned just how much I didn't know—which was a lot. But Sarah really dug my passion and drive. I was always the first one in the kitchen and the last one out. She recognized my strengths and weaknesses and my willingness to learn. She soon started to delegate major tasks to me.

One thing I really respected about Sarah was, no matter how slammed we were, no matter how deep in the weeds, she would jump on the line and bang with the rest of us. Every Friday night was jazz night; we offered a three-course dinner paired with wine and several canapés. I always worked the middle station, expediting and finishing off all plates with oils or garnishes. After a ticket (an order) came to the kitchen, I'd call it to the broiler, hot appetizer, or sauté stations (pantry and desserts got their own tickets). I had to be loud and concise. With conviction, I'd yell out, "Three New Yorks, two med, one rare, Juan; one bass, one scallop, Jared." And my cooks would immediately respond "Ordered, Chef."

My first cook, Juan Carlos, worked sauté to my left. On my right was an ex-con brother on the grill. He was a little weak, but he was banging hard not to fall behind. At about 8:00 P.M. one night, we got slammed hard with tickets coming in nonstop. Chef Sarah was on her way out after a twelve-hour shift, but she saw that we were going down. She tied up her hair and jumped in on broiler.

"CJ," she called to me, "give me an all day."

She wanted a rundown on all of the tickets that were up. "All day, Chef, twelve New Yorks—six medium rare, two well, four medium—five burgers, all with cheddar, seven lamb and four salmon." On top of these orders, I had to cook and plate a slew of first courses, including crab cakes, scallops, and soups and make sure the pantry (the station responsible for salads and or sandwiches) and dessert station were on point with the flow of the hot line.

The pressure was on. Waiters were all at the window waiting for their food. I stayed focused and tried to keep it all together. It was insane. I kept control of my crew, and forty minutes later the board was clear. The chef high-fived us, said great job, and left. She earned my full respect that night on the line. A lot of corporate chefs would never jump in and bail out the crew. Most would see all the tickets and go the other way.

Toward the end of our ninety-day takeover, Sarah offered me the sous-chef position at L'Escale. I was blown away, I felt on top of the world, but I knew deep inside that I wasn't experienced enough and the chef knew that as well. Still, she knew I had the tough street background to develop into a solid chef, and she was giving me that chance. Here was my opportunity to run a restaurant in a four-diamond resort just a couple of years out of prison. I told her I would discuss it with my wife and get back to her in the morning. Although I really wanted to take the job, I knew that my probation officer would have to approve it, and I had to make sure my wife was okay with me relocating to my old running grounds.

The main hang-up I had about relocating to San Diego was that so many people from my past would be just minutes away. This was a concern of Stacy's as well. But becoming a full-fledged chef so early in my career was an opportunity I couldn't turn down. I just had to convince my probation officer, and that's exactly what I did when I went to see her.

One thing my probation officer knew for sure was that I was true to being successful in the food world. Whenever I went to see her, it was all I talked about. I would go on and on about food and working in a kitchen, and I'd tell her that I was going to be one of the most successful ex-cons she ever managed. I knew that she believed in me, and it took almost no convincing for her to approve the transfer to San Diego.

After that, Stacy said, "Let's do it, baby" and we made the move.

I found us a little one-bedroom apartment at the end of the Silver Strand, less than five minutes from the Mexican border. The way I figured it, living all the way down there, no one would know I was back in Diego—no old homies and none of the women I dealt with in the past. I could stay totally focused. It would be just myself, my wife, my sister-in-law Sugar, who Stacy looked after like a daughter after their mother died a few years earlier, and little Jeffrey Jr. Sugar slept in the living room and Jr. slept with us.

Sarah and I began working on new menus. One of the areas where I lacked skills was in the administrative side. I had no knowledge of computers, P&L, writing reports, or doing employee reviews, but Stacy was good at all those things and I was grateful when she began teaching them to me.

Sarah and I ended up creating a dining experience that was unlike anything found at any other Marriott property. Marriott brass had given Sarah carte blanche in terms of keeping the high standards that preceded the La Meridian food and beverage division. We introduced foie gras; fresh fish was flown in daily from all over the world; and we bought the best seasonal produce available on the West Coast. After a year, however, we were notified by corporate that we'd have to mainstream our menu and product purchasing to conform to the Marriott status quo. As corporate

saw it, the restaurant needed to conform to all of its other operations, but we saw it as cost cutting and the end of the things that
we were most proud of in that restaurant. The bottom line had
won out.

They wanted us to remake our baby into just another mass-
market food factory. Sarah wanted no part of that. Neither did I.
As executive chef, Sarah was locked into a contract. I wasn't, and
I remembered everything Robert had told me about hotel chefs
who aren't given any creative freedom. I immediately started planning how to get back to L.A.

That's when things became strained between Sarah and me.
She was having problems with one of her cooks, so I told her I
wanted to take his spot in fine dining. I wanted to learn sauces
and how to cook luxury ingredients like foie gras, but Sarah said I
wasn't ready for it. She told me I couldn't even make a burger the
right way. When I challenged her, she put me to the test on the
broiler, and she was right. I got the wrong temperatures for the filet
mignon and the burgers.

"You're smart," she said, "but you need to slow down. It takes
years to master the techniques."

I knew that was the truth. I had yet to achieve the supreme
touch Robert had spoken of. I was still somewhat hardheaded
and cocky. Telling her that I wanted to go back to L.A. didn't
help matters. She lashed out about how she had given me a
chance. I felt guilty, but I couldn't stay there and become the
master of premade sauces and opening cans. I had to learn to
cook at a higher level—making everything from scratch, using
the finest ingredients, and so on—if I was going to be the high-
end chef I'd set out to be. I needed to learn from the best, not
just do what was easiest.

As it turned out, Marriott had just purchased the Ritz-Carlton
hotels and there was an opportunity opening up for a cook at the

Ritz in Marina del Rey. Although it was still another Marriott property, a three-star Michelin chef had arrived there and the restaurant was going to be among the best in the area. I sent them my resume and was accepted. I packed up the family and moved back to L.A.

When I got to the Ritz-Carlton in Marina del Rey, though, I wasn't given the job that I thought I had landed at The Dining Room. Instead, they had me doing breakfast and lunch at the Terrace restaurant. I was to cover the pantry station, as well as sauté and broiler, because management said I didn't have enough fine dining experience. True, but I was hoping for the opportunity to get that experience.

This was the first time I ever held a pantry position. The only salads I knew were prison salads and Robert's signature mimosa salad. I hit the Cook's Library, a cookbook store in Beverly Hills, and purchased a book on salads. Within days, I was making the salad specials and putting my own spin on the recipes from the book. In the meantime, I spent every lunch break in The Dining Room's kitchen watching the cooks, and trying to get the chef's attention.

Chef Gerard was an intense Frenchman, a real screamer who could be very nasty toward his crew if they didn't produce. That didn't bother me. After Robert, I didn't think there was any chef in the world who could rough me up. Hell, after prison, no kitchen intimidated me. Every day I asked him if there was anything I could do to help. He'd just look at me and shake his head no.

That didn't discourage me. I spent weeks making myself a

presence there. Finally, one of his prep boys called in sick one night and Chef Gerard was in the weeds. He offered to let me fill in. I was overwhelmed. I kissed his ass, saying "Chef, I truly appreciate this opportunity. Thank you, Chef."

My first task was to make Parmesan cheese tuiles for the Caesar salad. The chef then excused himself to go to a food and beverage meeting, and all the Latino cooks started eyeballing me. I was clean as a tack, had my A-game on. I quickly finished the twenty-five tuiles he asked me to make, and then cleaned out Chef Gerard's reach-in (the small refrigerator at his station). I wiped everything down and replaced all the sheet pans and put fresh parchment paper down on all the new ones. When he returned, Chef Gerard was very pleased with the tuiles and he noticed how I had cleaned and organized his reach-in.

I figured he probably didn't expect that from me. Black cooks had a bad rep in the food world. We were stereotyped as having anger problems, being lazy and combative, and cracking under pressure. Robert proved them wrong. Sterling proved them wrong. And I was going to prove them wrong, too.

After that night, the hotel's executive chef told me that Gerard wanted me to come work for him full-time.

"Thanks, Chef," I said. "That's the whole reason I wanted to come to the Ritz, for an opportunity to do fine dining with Chef Gerard."

He looked at me and said, "It's not going to be easy in that kitchen."

"You won't regret giving me the chance."

Before they would give me the job, though, Chef Mark wanted me to prepare a three-course tasting menu for him and a couple of the other Ritz chefs. I called Robert and asked for his consultation on the menu. I wanted to do a soup similar to his minted black beluga lentil soup with duck confit hash.

"Robert," I said, "I've got an opportunity to get into a fine dining restaurant, but I've never done a tasting on my own before."

"Just remember what we used to do here at the restaurant: Everything is tiny; focus on the flavor and presentation." Then he added, "Come down to the restaurant tonight when you get off and I'll help you out."

Later that evening I made my way down to Gadsby's. Together we made a small container of black beluga soup. This soup was the bomb. It is just black lentils slowly cooked in chicken stock with minced celery, onion, garlic, and torn fresh thyme and mint. It was a very light, aromatic soup and I was going to top it with some pieces of duck confit and serve it in an espresso cup.

That was one dish I wouldn't have to worry about preparing on the spot. For my second course I planned to serve a petit filet mignon with fingerling potatoes, carrots, and a black truffle jus (pan sauce). Robert suggested my third course.

"Remember the stir-fried wild berry dish with lemon Sauvignon sauce we served in the martini glass?

"Of course, Chef."

"Okay, this is what you do. I don't want you to take a chance of screwing up the Sauvignon. Get a little Grand Marnier, some plain yogurt. Mix them and sweeten it up with white sugar and a touch of cinnamon. You'll call that the cinnamon crème fraiche."

"Won't they know it's not really crème fraiche?" I asked.

"Okay," he said, and paused for a minute. "Don't risk it. Just call it cinnamon crème. Don't worry, they know you're an amateur. They just want to see that you can hold your own enough to cook with them. Just cook your ass off, make the food hot and well seasoned, and keep your timing on point. Dress the part and don't talk too much, just listen. They want Jeff the chef, not Jeff the hustler."

He knew my number. While my game was now cooking and my product was now food, I was still a hustler.

The tasting was at 3:00 the next afternoon. I was so nervous I was sweating, but I remained quiet and reserved. An hour before plating up, I began doing my prep and had a moment of self-talk. I hadn't had much experience cooking meat, and as Sarah proved, I still had a lot to learn.

I took my filets and seasoned them with kosher salt and cracked black pepper. I pan seared the meat and prepped the vegetables ahead of time and kept them at room temperature in case I fell behind while preparing the other items. The soup was in a pot on the stove, slowly simmering.

When the chefs came, I laid out the china and put my duck confit in the salamander to get it crispy. I spooned just enough of the hot soup into the espresso cups so each of them would have a few sips, then garnished each one with mint and the duck confit. I sent them out with complete confidence. All four cups came back empty.

With the filet mignon, I assumed they wanted it medium rare. I was ahead of the game since the filets had been preseared. Bringing a sauté pan to temperature, I drizzled in some olive oil and added a sprinkle of fresh garlic and shallots. Once the garlic and shallots sweated a little, I added some veal stock. While I let it reduce, I seasoned the potatoes and carrots that were simmering in another pan. Once the sauce was reduced, I finished it off with a little butter, and tasted it for flavor.

I scooped my vegetables and broth from the sauté pan into mini serving skillets, placed the filets on top, garnished with crispy leeks (which I had prepared earlier), and drizzled on a bit of rosemary-infused olive oil.

Again, all four dishes returned empty. I was riding on cloud nine. The adrenaline was pumping through my veins, the kitchen crew was watching, and I thought to myself in disbelief, *How can this guy put it down like that? How can I be this good?* I was so excited, I was clowning around. The only challenge now was to not overcook the berries. I got the pan superhot and dropped the berries in. I hit them with a tablespoon of white sugar and a splash of Grand Marnier. After the quick burst of fire to the flambé, the fruit was plated in seconds. I'd forgotten to add cinnamon to the yogurt, so I sprinkled it on top.

I waited patiently for the waiter to return with news.

He told me, "Henderson, Chef Mark would like to see you."

I took off my apron and toque and went out into the dining room.

"So how was everything?" I said.

"Great. We really enjoyed it. I'm moving you to fine dining on Monday. By the way," he said, "where did you get those miniskillets?"

"From Robert Gadsby," I told him. "My mentor."

"I like that you went the extra mile. You could have just used our china, but you brought in something new and different."

Monday afternoon was my first day at the Ritz-Carlton's fine dining restaurant, The Dining Room. It was very different from Gadsby's. Where Robert's place was chic and comfortable, this was very elegant and formal. I was nervous as hell and a little unsure of myself—I always felt like that when I was out of my comfort zone. Here was a great opportunity, so I knew that I couldn't punk out and be intimidated. I wouldn't be making any more money, but this was a chance to work in a real fine dining restaurant under a three-star Michelin chef. I showed up an hour and a half early wearing fresh clogs, new chef pants, and my company-issue coat.

I was expecting to get a copy of the menu and the recipes and be given the grand tour, but French chefs don't roll like that. Chef Gerard threw me right into the fire and watched me—how I cooked, how I moved, even how I taste-tested. I watched him as well, studied his moves, his style, and his personality.

Gerard started me on the pantry station. One of the first things I learned was the smoked salmon beggar's purse filled with Granny Smith apple crème fraiche and beluga caviar. Beggar's purses are a starter of a savory stuffing encased in a pastry like a little change purse. It was very challenging to tie the chive bows around the purse. I kept thinking about the supreme touch and focused intently on my hands. Several times, Chef Gerard would approach me and exclaim, "Like this! Like this!" when he saw me struggling with my heavy-ass hands.

The one thing I never taste-tested was the Caesar's dressing. I hated the taste of it so much that I had to have someone taste it for me—those anchovies didn't take well with me.

After a month, Gerard moved me to the meat station, and put me in charge of prepping and cleaning venison loins, foie gras, and whole beef tenderloins, and making stocks. Before long, I could tell how the meat was cooked just by touch and sight. But my main goal was to get to the sauté station because that's where most of the action happens during service. Sauté is the most prestigious station—all the most popular dishes come from there, like lobster risotto or loup de mer.

Tuesdays and Wednesdays were slow days and sometimes Gerard would leave early. He didn't work the line much anyway. Gerard always kept his menus and notes on the expediting side of the line. One night, he left and forgot his folder. After we cleaned the kitchen and everyone had gone home, I helped myself to quite a few menus and recipes and made copies of them.

I needed something extra to enhance my understanding of this unique man and his cooking. I knew that I had to understand him better if I was going to move up in this or any other kitchen. I needed to know how he developed menus, how he paired courses with wines, how he used seasonal products—things you don't learn on the line, and that most chefs won't teach you unless you are their boy, one of their top cooks. I wasn't going to be getting any private lessons in this kitchen; I had to learn in any way I could.

I knew it was wrong to snoop in another man's stuff, but Robert had once told me, "No one will ever give you anything, especially being as driven as you are. You have to *take* information from them."

Well, I'd never been to France, so I took the French experience from this French chef. I started to jack the CIA (Culinary Institute of America) schooling out of all the other chefs who had paid a fortune to go there. I'd quiz them, compliment them, whatever it took to get the information out of them.

"What did you put in there?" I'd ask. "Damn, how'd you come up with that?" I stroked them good. I borrowed their books to learn their lingo and I sometimes tape-recorded the preshift talks. Even to this day, my pronunciation of certain terms isn't that great; after all, I had been speaking *street* for more than thirty years. But my street talk in the kitchen sometimes allowed me to excel with the Latino line cooks and the other immigrants because we all understood where we came from. They mostly just wanted a paycheck; I wanted knowledge and information to get to the top.

Chef Gerard never spoke to me about anything other than cooking for the first few months, but eventually he warmed up to me. Once he took me aside and showed me a photo of himself fishing in the Mediterranean with about a dozen other famous chefs. He told me it was from back when he had celebrated being the youngest chef ever to earn a three-star Michelin ranking.

One of my fondest memories was the night we prepared a VIP dinner to honor French chef Paul Bocuse. Joël Antunes from the Ritz in Buckhead, Atlanta, was there as well, and I was paired up with him on the night of the event. Chef Antunes cooked in Europe before coming to Altanta and was trained in classical French cuisine, but he had a light style and a Thai influence. He was cool, laid-back, and we understood each other.

That night Joël and I really got into it, banging out the lobster risotto course. I had done most of the prep before the other chefs arrived, cooked and cleaned all the three-pound lobsters, organized the stations and stocked the kitchen, so when the big boys arrived all they had to do was cook. At the end of the dinner, Joël hugged me and said, "You did a great job. If you ever come to Atlanta, you've got a job."

Just a week later we were notified that The Dining Room would be closing down and turned into a banquet space. We were livid. It came down to money. Chef Gerard and his staff cost the hotel more than it could possibly make off the thirty covers we did a night. The company decided that having a fine dining restaurant wasn't worth it, and so it was shut down.

I didn't want to go back to the Terrace or do banquets, so I decided to move on. My mom was nervous that I kept changing jobs. She didn't understand that to be a well-rounded chef you have to expose yourself to all the different aspects of high-end cuisine. I assured her that I knew what I was doing and reminded her that I wasn't a young buck anymore and that I had to get mine now. I didn't attend the Culinary Institute of America like the others, so I had to hustle in other ways. Stacy was down; she understood and saw the future. Besides, I wasn't selling dope, so whatever I wanted to do was cool as long as it would further my dream of becoming a top chef.

I went by Gadsby's and told Robert I wanted a new job. Robert said my timing was perfect since he was consulting on the opening of a club/restaurant called The Good Bar and offered to bring me in as his sous-chef for $15 an hour. It would be just for a few months, but it would be a good experience.

The Good Bar was off the hook from the night it opened. Celebrities and athletes were coming through all the time. The kitchen was fast paced and Robert was training me on butchering and sauces. For the grand opening, I worked pasta and the broiler, and Robert was on sauté and expediting. We were banging on the line. Robert saw that I had developed a much better touch with the food. One night, though, he saw me rough up the waitstaff. They were all Hollywood types, just there for the money and with no respect for the kitchen crew. All night they had been messing up and having me refire dishes until I finally lost my temper. I started yelling at them and intimidating them with hard looks. Robert had taught me never to let the waiters run the kitchen and not to lose control. But he was disappointed that I had lost my cool, and so was I. The next day, I apologized to all of them, and it never happened again.

After two months, Robert's contract to get the place up and running expired and I had to move on once again.

My next tour, I teamed back up with Sterling Burpee, the chef who got me my first interview at the Marriott. Sterling had opened a new restaurant in Westwood called Zing. This was a different type of place. Comfort food: steaks, fish, pastas, bar food, great soups, and weekend brunch.

From time to time the owners started lagging on pay. Sometimes we wouldn't get a check for two weeks. That just wasn't going to fly. I had a son and Stacy was pregnant again with baby number two. I thought I had enough on my shoulders with my son, my sister-in-law, and my wife all under one roof. I expressed my concerns that we weren't ready for another child.

Stacy asked me, "What are you saying, Jeffrey?"

"You know I'm focused on my career," I said. "I feel like we're doing things backwards. White people make their money first and then they make a family. This is going to make it harder to get where we want to be."

But Stacy wanted a big family. I didn't mind that, but I wanted to be able to provide for them—pool, bikes, nice clothes, tutors. The whole nine months I wasn't very supportive. All that changed when my daughter Noel was born. She was like my twin.

I was a little apprehensive about telling Chef Sterling that I was ready to move, but I had no choice. Zing just wasn't my style, it wouldn't get me any further along, and I just wasn't confident that I'd get every paycheck that I was due. He'd be pissed because he depended on me, but I had to think of my own career. Sometimes I felt caught between him and Robert, my two teachers, and my own goals. I always dreamed of the three of us taking on the food world together. Robert would never have that, though. He was the Lone Ranger, always had to be the man.

When I told Robert that I needed another hotel job because my wife was pregnant again, he said he would make a couple of phone calls and would introduce me to Gary Clauson, the executive chef of the Hotel Bel-Air.

A few nights later, we jumped in Robert's BMW and headed up Stone Canyon Road to the Bel-Air. It was a five-star, five-diamond, old Mediterranean-style hotel, with just ninety-nine rooms. Nancy Reagan ate there every Wednesday. Oprah, Michael Jackson, Robert De Niro, and Madonna were all regulars, plus the occasional knucklehead rappers.

We met Gary at Table One, a private dining room in the kitchen where, for $150 per person, a five-course meal was cooked right before your eyes.

"This is Jeff," Robert told Gary. "He's a very talented young man and he's always wanted to be a chef." He went on to praise just about everything about me, and Gary gave me an application to fill out. Several weeks later I accepted a position as chef tournot at one of the country's most prestigious restaurants.

Since I first started working in restaurants, I had worked long hours—ten- or even twelve-hour days—and spent all my time off hanging around Gadsby's restaurant or with my face buried in a cookbook. This job would be no less consuming, plus I'd have a forty-five-minute drive each way, which meant that I would have even less time with my family, but they supported me and believed in me, and that helped keep me focus. This faith would be something that I relied on as I faced this next restaurant and my toughest challenge as a cook.

FOURTEEN

※

SABOTAGE

On my first day at the Hotel Bel-Air, I reported to Tom Hanson, a six-foot-three-inch white guy, the executive sous-chef, and the number two man in the kitchen. Chef Hanson was very talented, a great cook, and a master of presentation. He was an out-of-the-box chef, an original, who always worked the line with his crew. Tom was the type of chef who didn't mind getting dirty—my kind of chef.

The kitchen was banging and my stomach was butterflying. I was worried about whether I would be able to hang with this new crew. Though I'd worked at some top places before, the Bel-Air was the crème de la crème of hotel restaurants in L.A. I knew my A-game would have to jump to an A+ if I was going to make it here.

Gary was a no-nonsense chef, notorious for reaming cooks and other chefs for not taking care of their business. They're always nice to you at the interview; they never tell you the truth about what really goes on in the kitchen because they want to paint a rosy picture for the fresh blood coming in.

After the obligatory grand tour, Chef Clauson introduced me

to the crew. The first kitchen soldier was Mario. The most power-ful and talented member of Gary's staff, Mario was a tough Mexi-can guy, the kitchen's saucier and the man in charge of the sauté station.

Feliciano was number two under Mario, a stocky San Sal-vadoran who ran hot apps, the middle station in the kitchen. Feliciano never smiled, and I knew I'd have to work to earn his and Mario's respect. No prison-style intimidation tactics were going to have any effect on these guys. This was clearly going to be the most challenging crew that I had ever had to prove myself worthy of.

Another tough guy was a sous-chef named Umberto who had a strong street rep and was tatted up all over his body. From the moment I started, he never so much as looked me in the eye, so I knew we weren't going to hit it off. He controlled the prep crew in the back, and those dudes were like robots. They never even spoke—all they did was prep, prep, prep from the time they walked in until the time they left. They made all the stocks and the employee meals, and they did all the butchering. They were very talented and focused, but always in the trenches, overwhelmed with work.

There were also two women in the kitchen: a very talented Asian woman named Arlene, who'd graduated from one of the top culinary academies, and a Spanish girl named Maria who was beau-tiful but lazy. Women were accepted a lot more quickly than men because there isn't that macho competitive vibe when it comes to females. Arlene was a professional, but Maria was fucking one of the other cooks. She also used her feminine charms to manipulate all the other guys into doing her prep work and pulling her out of the weeds during service.

The vibe at the Hotel Bel-Air was different from any I'd ever been around—even more competitive and professional. And, the place itself was different, too. During my tour of the kitchen, I asked where the walk-in freezers were.

"We don't use them," Chef Hanson explained. "This is five-star fine dining. Everything we do here is fresh. All our meats, sauces, stocks—nothing comes out of a can and nothing goes in a freezer. We ship everything in fresh every day and use excess for staff meals."

A big change from the Marriott standard operational procedures mentality.

The chef tournot is a jack-of-all-trades; he can work every station from broiler to hot apps, sauté, pantry, and saucier. He's the go-to guy. The head chef depends on the tournot, especially when the kitchen is down a man.

I was a long way from excelling as a tournot since I still had so much to learn. If I didn't know how to do something, I would pretend that I did. I'd sneak off to the walk-in with a cell phone to call Robert in the middle of service and ask him how to perform particular tasks. It brought me back to my first few years in prison, when I'd been told so much about white people being superior because they had all the knowledge. I remembered the teachings of the black prison scholars I studied, who wrote that if anyone wanted to keep knowledge from a black man, all you had to do was put it in a book, because we didn't read. "If you want to know what the white man knows," they had preached, "read his books."

Robert had more than a thousand cookbooks so I started adding to my personal library after my first day at the Bel-Air. I bought *The Sauce Bible,* a book called simply *Foie Gras, The Food Lover's Companion,* and the latest edition of *The Professional Cook.* These books covered many of the basics of the cuisine prepared in the Bel-Air kitchen. Every night I spent hours after work studying those texts and honing my skills.

The one skill you couldn't learn from any book was how to successfully maneuver through the cutthroat politics of a top-flight kitchen. The trouble started one Friday night when I was paired with Maria at the broiler station. That station took the most heat during service, because that's where the majority of the Bel-Air's popular dishes were prepared, like rack of lamb, grilled pheasant and other game, and even some fish. We had to perform above the rest of the crew, just to hang with the flow of service. Maria was my partner on the station and her game was weak. I was cooking the meat while Maria was supposed to be preparing and plating the vegetables and starch. She was fucking up from the get-go. I had a rack of lamb ready for Table One, the table in the kitchen, and Maria was just standing there stressing and staring back and forth between me and an empty plate. It kept going like this until I was so caught up trying to get her to do her job that I misheard the chef when he ordered another rack of lamb.

He was yelling back, "Mario, let's go with the twelve top! Pick up! Jeff! Maria! Let's go!"

I said, "Fuck. I'm down a rack." We were short one portion of lamb that I'd have to prepare from raw to plated in minutes.

I couldn't let Chef Clauson know that I'd fallen behind, so I ran to Mario's station and grabbed one of his hot sauté pans, which the dishwashers always made sure he had plenty of while I was always short clean pans. After quickly seasoning up a rack, I mashed it down and put it straight in a hot pan.

By now, Chef Clauson was in a screaming rage. "Jeff, I want my fucking lamb now!"

"Lamb's coming, Chef! I'm down one rack. Let's send out everything and by the time the waiter gets back I'll have that lamb dish plated."

The chef was going crazy, yelling and cursing his head off. I was sweating bullets and Maria was ready to collapse on the floor—it was killing her. It was killing *me*! I felt weak, ashamed that I was letting the chef down and causing the rest of the crew to get backed up. I knew I had to put my gangster face on and take full control of my station. Failure was not an option.

I remembered what Robert had always told me: "speed, taste, and presentation." And, "Never depend on another cook to back you on your station."

Even after that last rack was plated and the service ended for the night, I couldn't face Chef Clauson. Then I came to learn something about him, which I'd later learn was typical of most top chefs. They'd whup your ass during service, get in your face, throw things, the whole nine yards, but at the end of the day, most of them never took it personally.

When this night was over, he came up to me, patted me on the back, and said, "Good job, Jeff. You're going to be a good soldier."

Damn, I said to myself. What do you mean "good job"? I fucked up. This can never happen again. That girl has to get off my station. I've got to get rid of her.

I knew there would be consequences. I'd have to deal with her boyfriend in the kitchen, not to mention the rest of the crew. Even the waiters loved her. The ones who didn't want to fuck Maria thought of her as their little Latin sister.

The next day I came in wearing my soldier's face. Like always, I was there an hour early to make sure I got my prep work done. When Maria came in, I immediately started giving her instructions. She looked at me like I was crazy. She'd been there longer, so who the fuck was I to be giving her orders? I didn't give a shit: I was not going to be embarrassed again. I was taking full responsibility for the success or failure of the broiler station.

"I'm tired of your shit," I told her. "I don't give a shit how good you look. Your boyfriend won't be able to do shit for you if you make me look bad again."

Even though the boys on the line loved her, they couldn't stand up against me because I was bigger and tougher than any of them, and they knew it. Her boyfriend didn't want any trouble; he was between a rock and a hard place because his team needed me. I could tell that most of the crew didn't particularly care for having a *negrito* in the kitchen—on top of that a strong one. They also knew I was no easy pushover, and I would be a better asset in the kitchen than Maria. I knew Mario at least respected me professionally, because he would watch me prep and could see my speed and organization was top-notch. Umberto watched me, too, but he despised everything I did. As he studied me that night, I studied him as well, watching his every move.

I started my prep by searing off two or three extra racks of lamb and wrapping them in foil. I hid them under my little chrome utility cart on the side of my six-burner stove so that, if I fell behind again, I'd have backups that I could pop in the oven and bring up to temperature. Sandbagging food that way was definitely against the rules, but I felt I had no choice. It was crazy in there, it was a battlefield, and I intended to win by any means necessary. Within half an hour of service, the chefs would be ranting and raving, waiters stressed out, and all of these Hollywood celebrities watching the show from Table One. Even as I prepped, a part of me felt like waving a white flag and saying, "I give up! I'm burnt!" And then just walking out.

We were an hour into the service when I got my first lamb order. I seasoned the rack of lamb with my kosher salt and cracked black pepper, seared it off in a hot sauté pan on both sides, and put it in the oven. I told Maria to drop the polenta in the deep fryer and in the next twelve minutes to sauté the spinach and to caramelize a shallot. My lamb came out; I cut the rack in half and let the pieces rest a few minutes before setting them over the crispy polenta with the double bones crisscrossed and pointing upward. Then I spoon-drizzled the red wine sauce around the chops, garnished them with crispy basil, and placed the dish in window.

Chef Clauson told me, "Great presentation, Jeff," but within minutes it came back from the customer.

"What's wrong with the lamb?" Chef Clauson asked the waiter.

"They didn't like it," he reported. "They said it was kind of sweet or something."

The chef told me to refire it, and I could already feel all eyes on me. I noticed some of the boys at the end of the line were smirking at me. For some reason, a sixth sense told me to check my seasoning. I'd heard that the more competitive it gets in high-end kitchens, the more cooks sometimes tamper with their in-house rivals' seasonings and sauces—adding water or soy sauce to their sauces and red wine, for example—things you couldn't fix on the fly. So I tapped the tip of my pinky to my tongue, touched my kosher salt, and tapped it on my tongue again—

The *motherfuckers*! Someone had mixed granulated sugar with my salt.

Immediately, I called to Chef Clauson, "Come look at this bullshit, Chef. Someone put fucking sugar in my salt. I've never come snitching to you but this is over the top. It's gonna get out of hand and someone's gonna get hurt. I appreciate the opportunity you're giving me, Chef, but I was treated with more respect when I was in prison."

I let him know that I was a professional, not just some ex-con, and that I could run my station well if he could get his crew to stop interfering with me. I had, after all, run kitchens over some of the world's most desperate men.

He ended up just letting it slide. I was furious. I didn't need any more proof that the boys wanted me out.

The dishwashers had already stopped supplying me with enough sauté pans and, two weeks before, all of my prep had been thrown out and I'd had to start from scratch. I didn't bitch about it any further to Chef Clauson, but I did get in Tom Hanson's ear a little, explaining that someone was trying to set me up. As far as the chefs saw it, though, the drive and passion I brought to the kitchen was by no means worth their having a conflict with the Hispanic guys on the crew. They simply represented a greater asset than one lone ranger. I knew I'd have to deal with them on my own. And that's just what I set out to do.

I started to flex my muscle on the line, mad-dogging them, staring them down. In my own way, I let them know that I wasn't the one to play with. I further let them know that if we couldn't deal with the situation there in the kitchen, we could deal with it on the street. At that point, I was even ready to bring in some boys from my old days to deal with these fools in the parking lot. The motherfuckers had backed me into a corner by playing dirty. The way I saw it, anything that happened to them now was something they brought upon themselves.

If my career was going to get fucked, so was everyone else's. But, before things went that far, I thought I'd try a little diplomacy—like how the corporate gangsters do it.

Of all the Latino guys, Mario seemed to me the easiest one to flip. When I say "flip," I mean play on his intelligence and manipulate him psychologically until he would let down his guard and accept me as a true member of the crew. I felt I could work my way in with Mario because, when he wasn't in the presence of the rest of his boys, he would share some of his insights and knowledge with me. It was always just enough for me to get by, though, never enough to build my confidence.

At the same time, I started a crew-wide public relations campaign by toning down my aggressive attitude and extending the olive branch by letting the crew bitch me up a little bit. They had me doing prep work that wasn't part of my job, and I let them. Mostly, though, my focus was on Mario. I started coming in earlier and earlier to help him get his fourteen sauces ready for service, set up all of his pans and reduction pots, and put out his wines and the various ingredients he needed for service.

The sauce that required the most prep work was the mango sauce that was paired with the duck. I peeled the mangos, removed the flesh from the pit, chopped up the cilantro, reduced the coconut milk, and pounded the essence out of the lemon grass. I made Mario's job so much easier that, despite his playing Mexican mafia games with me, he started to respect me. Before long, he began to share with me his secrets about the sauté station and making sauces.

Just like with Robert Gadsby and Sarah Bowman, the things Mario taught me in action couldn't be learned from any text or in a culinary school.

After the white boys in prison told me I was intelligent and the brothers hipped me to self-help books, I knew that I had as much a chance as any man to become successful. I was learning that no top chef is superhuman. I saw them digging through books, poring over magazines, and stealing other people's creations. They were jackers just like T-Row and me. The only difference was the product. When T and I were out jacking cars, these guys were jacking recipes, formulas, and techniques.

Once I mastered the broiler station, Chefs Gary and Tom knew that I was going to make it. They recognized that I was a soldier and not some wannabe. With those two behind me, the rest of the Bel-Air crew grudgingly started to give me their respect.

After four months, Tom told me I would be filling in for Mario on his days off. I was happy as hell. For two days a week I would be working the most coveted spot on the line. This was my ultimate opportunity to shine and also hone my sauté and saucier skills.

I'd been studying Mario from day one—his technique, his every move; how he arranged the pots on the stove; how every handle for all fourteen sauce pots always pointed in the same direction; and how every one of those handles was labeled with duct tape indicating which sauce was which.

Within a month, I had the sauté station down to a science. I became so confident that I would even jump from the broiler to sauté when Mario was slammed to give him a hand. I had Mario just where I wanted him. It wouldn't be long, I knew, before the rest of his boys fell in line completely.

Although Feliciano was still a little cold toward me, he didn't make any problems either. Soon enough, the only person in that kitchen who I had any real trouble with was the one person who'd been giving me grief from the start: Maria. She wouldn't have been a major problem, but she had one advantage over me—she was still fucking that other cook and their relationship was starting to get serious. Even after the rest of the crew had accepted me, she still resented me for forcing her to do her job.

She brought up fictitious charges against me with human resources, claiming I had threatened her, so a hearing was to be held where I would have to defend myself. Sure, I yelled at her and barked some orders, but that happens in kitchens everywhere every day. Hell, I never worked for anyone who didn't get on my ass when they thought they had to. Still, I was a large, intimidating black man, an ex-con, and a felon. And she was a very pretty and petite young woman who most of the front and back of the house staff wanted to bed. I didn't like my odds.

At the hearing Gary Clauson came to my defense. When Gary spoke, he sounded like a top-notch defense attorney. I barely had to say a word. For the first time, I felt like I really was a valued member of his team. The charges against me were dropped. After that, even Feliciano was down with me, and the job became easier, less stressful, and more enjoyable.

Over the next couple of months I made another new ally. Josh Thompson was a young chef out of New York City who oversaw the tasting menus for Table One. He had worked under Thomas Keller for two years at the French Laundry and for Paul Bocuse at L'Auberge du Pont in France. Josh's dishes were light and seasonal with remarkably bold flavors and sexy presentations. Josh and I had very similar philosophies about cooking. Maybe that's why he decided to take me into his confidence and started teaching me things. Of course, I always wondered if it was because he wanted to see me grow as a chef and was grooming me to become one of his soldiers, or if he simply wanted to bitch me into doing his prep work.

I didn't really care, because Table One was the ultimate. I wanted to have an influence there. A lot of high-powered African Americans ate at Table One, so I thought I would suggest something southern, something that played on the flavors that most blacks grew up loving, but I was too afraid of being rejected to make the move. After a year, I finally got the break I'd been looking for: Josh had a hernia operation.

Gary and Tom started covering Table One while Josh was recovering from surgery. Gary didn't have a lot of energy. He had been recently diagnosis with leukemia. And Tom didn't have the time and didn't want to be stuck with another responsibility. So I asked Gary if I could try my hand at writing a tasting menu for Table One and execute the tasting.

Gary decided to give me a shot, so I wrote up a four-course tasting menu. It was summertime and peaches were in full bloom, so for the first course I went with a strawberry and doughnut-peach soup with a hint of elderflower syrup. The appetizer was a pan-seared Muscovy duck breast with braised Napa cabbage, caramelized sweet potatoes, and a port wine gastrique, and garnished with crispy leeks. For the main course I chose beef Rossini, a classic dish of filet mignon and foie gras. The finish was pan-seared loup de mer with braised Swiss chard, carrots, and tarragon cream.

Once Gary gave me the go-ahead, all the waiters were, like, "How are they gonna let this guy do a tasting for Table One? He's not even a chef!"

The first course was easy. I took pureed strawberries and then added a little bit of fresh squeezed orange juice to thin them out and a touch of sugar to enhance the sweetness. I strained the mixture through a china cap to remove all the little seeds. As I watched it come through the strainer it was like pure strawberry juice. After a quick whisk of the juice, I drizzled in some elderflower syrup to intensify the flavor, poured the juice into a white soup bowl, peeled and sliced a doughnut peach, and folded in fresh minced peppermint. Then I set that aside for the waiters to taste so that they could talk it up to the customers during service.

For the second course, I cut a four-ounce piece of duck breast and scored it with a sharp knife to keep it from buckling and to allow the seasoning to flow through the skin when I seared it. After searing it off in a sauté pan, I placed the duck in the oven at a slow roast. I took some Napa cabbage and caramelized it in a pan of its own. I did the same with the diced sweet potatoes, added brown sugar for the sweetness and the maple flavor, fresh thyme and butter, and gave it a quick sauté until the edges of the potatoes were dark.

On a square china plate, I arranged the cabbage and sweet potatoes in a little mound. Then I removed the duck breast from the oven and let it rest for a couple of minutes before cutting it into four thin, medium-rare slices. These I placed over the vegetables, then spoon-drizzled the red wine reduction around the artfully arranged dish and garnished it with a poached bing cherry.

Next came the Rossini. I seared off a four-ounce beef filet and scored a small piece of Hudson Valley foie gras, which I would sear at the last moment possible because it's as delicate as butter. Rob had taught me to sear filet mignon from the sides, the top, and the bottom, and to sprinkle my seasonings from high up with long, sweeping motions like an artist's paintbrush strokes in order to cover it everywhere. Once the filet was done I plated it, and immediately put the foie gras in the blazing hot pan, seasoning it with black pepper and sea salt. In less than a minute, it was sitting beautifully atop the filet.

Finally, I seasoned up the fish and, when I laid it in the hot sauté pan, I could smell the fresh thyme, fresh cracked black pepper, and salt filling the air as the fish began to sear. I love the sound of fish crackling in a pan—it's like it's talking to you. It speaks to your senses and arouses your palate. When I'd finished it on top, I set it aside to rest while I a quickly sautéed the Swiss chard and fingerling potatoes.

When I plated that last dish, it towered over all the others. Rob had taught me how to stack food, which was a technique pioneered by Alfred Portale at Gotham Bar and Grill in New York. It was beautiful, all the colors playing into one another like a rainbow.

The waiters were blown away. The crew was blown away. Gary put his hands on his hips and stared down at the food and there were no objections, but one. He told me that strawberry doughnut-peach soup was a little too sweet to be served as a first course. I took his criticism with pride. I felt unstoppable, but I'd still never let my guard down.

Shortly after my first success at Table One, word went around that the hotel's banquet chef was about to quit. Even though banquet chefs got very little respect, it was a chance for me to make history. No African American chef had ever been in charge of a kitchen at the Hotel Bel-Air. I applied for the job and Chef Gary told me there was one candidate above me because of his long experience in the industry.

I didn't take it as a no but, instead, as a consideration—and I was pleased. As it turned out, the more experienced candidate didn't last three weeks before the Latino boys ran his ass out of there. I was offered the position at $28,000 a year, plus benefits.

It wasn't much, but I was more concerned with the title than the pay. But after Stacy did the math, she said, "Jeff, this doesn't make sense. You'd make more money as chef tournot doing overtime than you will as a chef."

I hadn't looked at it from that perspective; I'd only kept in mind what Robert had told me, which was that there were two forms of money: cash and experience.

"If you go for the cash first," he'd said, "before you get the experience and exposure, you'll burn out before you have the experience to make it all the way to the top."

So I took the experience, telling Stacy that the money would have to wait.

Within two months on the job I was getting down. I was banging out high-end parties for a hundred people at a time, and everything was made from scratch. Chef Clauson let me be creative with the food I served, so long as there were no shortcuts taken. Some of my best dishes were diver scallops with applewood smoked bacon and pencil aspargus wrapped with smoked salmon. It was intense at times, but I was confident and focused enough to flourish. I was also confident enough to bring up the subject of my pay with Chef Clauson.

I let him know that I had been told that my position normally paid $40,000 a year and that I wanted to be compensated at the same rate as all the chefs before me. He agreed and gave me the raise I was asking for. That's when I was issued my first legitimate chef's jacket. It read "Hotel Bel-Air" on one front side and "Jeffrey Henderson, Sous-Chef " on the other.

I wore that jacket with pride, honor, and respect—and I wore it everywhere I went. Even when my shift was over, I wore it on the street, when I went to the store, the gas station, anywhere. People in my community would recognize me and ask about my job and cooking. I thought back to being in prison and telling my family and Mr. Hershman that I was going to be a world-class chef someday.

If I wasn't on my way now, I didn't know what "on my way" meant.

Six months later, I was being recruited by Joseph Antonishek who, at just twenty-eight years old, was the executive chef and food and beverage director of the five-star, five-diamond L'Ermitage Hotel in Beverly Hills. I knew it was time to move on. I had conquered the title of chef tournot, learned every station, and had made history by becoming the first African American banquet chef at the Hotel Bel-Air. I'd earned the respect of Gary Clauson and the crew, and I had even made Robert proud.

My resignation loomed heavily for several weeks as I contemplated leaving. Chef Joseph was calling almost every day. I couldn't face Chef Clauson, so I went to Tom. He was angry but he understood. When he consulted with Gary, he understood as well. They knew I had to keep moving forward to expand my culinary career and seek out new challenges.

After a couple of spy missions to L'Ermitage—snooping around, asking questions about Chef Joseph—I was a little hesitant because I heard that cooks had a hard time lasting long there. But I was confident that I could make it through anything, so I gave my official two-week notice at the Hotel Bel-Air. On my last day, I shook hands and hugged everyone, thanking them all for their patience and what they taught me.

L'Ermitage (now Raffles L'Ermitage) is located in the center of Beverly Hills with 129 incredibly expensive suites and a cool rooftop banquet space where rappers throw record release parties. My first assignment was the lunch and dinner shifts. I was hired as a sous-chef at $42,000 a year.

Since I didn't have a name in the industry, I never understood why Chef Joseph had pursued me so eagerly and he didn't even ask me to prepare a tasting, but I soon learned that his other soldiers were falling off. Two sous-chefs had quit, one of them to work at the French Laundry. Before long, the only full-fledged chefs left were Joseph and myself. I ended up doing breakfast, lunch, and dinner. In return, Joseph promised to teach me the financial end of the business, where I lacked knowledge. Robert once warned me that very few chefs would teach me that, because once you could cook, manage, and understand profit/loss sheets, you could replace anyone.

Joseph never did hold up his end of the bargain, but my food took off to another level. Joseph's food was light and Asian-inspired, similar to Robert Gadsby's. There was, however, nothing intimidating about him. When he yelled at the cooks, no one even twitched. At that time, he lacked the maturity to marshal his forces in the kitchen. By using my advantage of being older, even I influenced him to do things my way.

I had only been at L'Ermitage a short time when Joseph left. That left me as the only chef running L'Ermitage. Just over three years out of prison and I was overseeing a five-star, five-diamond property in Beverly Hills. The pressure was on. And there was trouble ahead.

The general manager promised me advancement in the company if I held down the fort until they found a replacement for Joseph. I agreed. For some reason, I really thought I could run the place smoothly and become the permanent executive chef. Little did I know that there was a revolution brewing in the front of the house. There was a conspiracy to undermine me and run me out of there with my tail between my legs.

Not yet 90 percent polished, not yet 100 percent distanced from the street game, I allowed the provocateurs to get under my skin and push my buttons.

It was Saturday night service. The restaurant was packed. The manager of private dining started to fuck with me by intentionally delaying ticket times, by not putting them into microsystems so the cooks could prepare the orders; those orders piled up and waiters began screaming, "How much time for table eight?" and "Room service!" My cooks couldn't handle it and started dragging behind.

I should have listened to the rumors about L'Ermitage's high turnover rate for chefs and cooks. Now I was seeing it firsthand: Behind the cloak of a prestigious restaurant operation and hotel, it was completely unorganized and unprofessional.

If I had Mario and Feliciano down here, I told myself, we would be banging this out with no problem.

My anger rose as they kept pushing on me, my blood pressure surging. Finally, I lost it. I went off on both of the managers and whispered to one of them, "I'm gonna fuck you up for this."

I meant that I was going to report her to the GM for her sabotage in the kitchen. She took it as an opportunity to say that I was going to physically harm her. She screamed out that I was going to kill her.

Within minutes, security showed up. "Chef Henderson," the guard said. "You need to grab your things and leave the property."

Meanwhile, the girl was sweating and turning red like a beet. She should have won a fucking Oscar for her one-woman show. I grabbed my knife bag, got in my car, and took off for home—nervous as hell. I was still on probation and didn't want any drama. I never went back to L'Ermitage. (I never even called or got my last check.)

It was a long ride home to Harbor City. When I looked Stacy in her eyes, she knew something was wrong.

"What did you do?" she asked me.

"I fucked up. I talked crazy to that white girl I've been having trouble with at work."

"Baby, don't worry, you'll get another job, but you have to work on your anger."

"I don't have any fucking anger problem!" I shouted. "I'm just tired of these motherfuckers fucking with me and trying to stop my progress."

"But, honey, it's got to come to a point where you deal with it in another way. You can't keep going off on people because it's gonna end up with you not being able to find work in the kind of places you aspire to."

My hard head reared up once again. "I'm not worried about

that," I told my wife. "I don't believe I have an anger problem. I'm just as angry as anyone else in America. People fuck with you, you're gonna get mad."

"But, Jeffrey, never forget something: You are black, you are a convicted felon, and you are intimidating to people who don't get you. You've gotten jobs because you speak well and know how to play the corporate game. But you can't show your street side because they will hang you every time."

"I understand that. But at the end of the day, I am who I am. I get fed up not being who I am. I'm tired of living a double life and being fake to learn the cooking game. Sometimes I just lose it."

"You can't lose it," she said, "at the expense of your family and your career."

"Yeah," I said. "I know that, baby."

FIFTEEN

CHEF OF THE YEAR

Weeks went by without any new career prospects, and I began to worry about my ability to get another job like the ones I'd had. I'd worked at some of the best restaurants in Los Angeles, and now my temper had put an end to all that.

I had been hearing a lot about Las Vegas. New hotels were popping up all over the desert city and I was told that they were always hiring in the service industry. The pay was said to be better than in L.A., too. I was almost done with my five-year probation, and my probation officer thought that Vegas might be a good move. I received permission to go there for three days to check out prospects.

Stacy was wary of the idea. She had been following me around Southern California ever since I'd gotten out of prison and, even though she believed in me, she was worried about the lingering issues that still weighed heavy on my mind from my years of incarceration—the anger, the claustrophobia, the paranoia. Her fear was that I'd end up losing control in a place where we didn't know anyone or have any friends.

I finally convinced her to say okay after I told her about the great potential for housing and wages that Las Vegas promised. We were low on funds but she knew that a plane ticket to Vegas was an important investment. "Honey," she said, "do what you need to do."

Money was especially tight because our second child, Noel, had been born six months earlier. Before I made the trip to Vegas, Stacy and I did our homework and faxed my resume to every hotel on the strip. I began writing tasting menus and calling cooks I knew who lived in Las Vegas to get as much info as possible about the top chefs and high-end hotel movers and shakers. My target goal was the Bellagio. The Bentley of Las Vegas resorts, it was the only five-diamond (the highest rating for a hotel) property on the strip, and I wanted to be part of it.

I checked into the Jockey Club, an all-suite budget hotel next door to the Bellagio, on a Sunday morning. My plan was to do recon that afternoon before beginning my employment search the next day. I hit the Bellagio and hung around the employee entrance on Frank Sinatra Boulevard. I was on the lookout for black cooks to sound out about the executive chefs—their personalities, styles, and tastes. It was an hour before I spotted a black cook.

Damn, I thought, *this place is no different than L.A.* No *negritos,* just Latinos and a few white guys.

Approaching the brother, I went into my usual spiel: "Hey, man, my name is Jeff Henderson. Can I talk to you for a minute?" I asked him who the executive chef was, explaining, "I'm a chef from L.A. and I've always wanted to work here at the Bellagio. How can I get in the door?"

The cook was very cool with me. He said, "There's just one person you have to go through. Grant McPherson is the most powerful executive chef in Vegas."

I'd heard of him. Grant was supposed to be a hard-nosed chef who accepted nothing less than perfection and expected 100 percent from his culinary brigade.

"Is that right?" I said. "Let me ask you: Are there any black chefs here?"

"No, man. No, there's not. This is Vegas."

"Oh, boy," I thought.

After he'd gone, I went to the front of the Bellagio and checked it out further, gathering information in case I scored an interview. That same afternoon I cased Mandalay Bay, the Venetian, Paris, Aladdin, Treasure Island, the Luxor, and the Mirage.

First thing Monday, I called every hotel I had faxed. The Luxor called me back within an hour and gave me an appointment for early that same afternoon with the executive chef. I called Stacy, psyched that things were happening so quickly, but my excitement dimmed as soon as I showed up for the interview.

The executive chef was an overweight German guy and his kitchen was depressingly unorganized. I knew right away that I didn't want to work there and I went through the interview half-heartedly. In the end, I told him I had other interviews that afternoon and that I'd get back to him about doing a tasting.

At the Aladdin, the executive chef showed me his kitchen and then scheduled me to do a tasting that very day at 4:00. Back in my room, I opened my laptop and wrote a menu based on what I'd seen in the kitchen of Tremezzo, the Aladdin's northern Italian restaurant. I wrote up five courses, simple and light, and was all ready to go. I put on my chef gear, grabbed my knife kit, and headed out. The executive chef gave me three hours to prepare, saying I could use anything from his walk-ins or off the line.

As I set up shop in the prep kitchen, one black waiter named Kenny seemed very interested in what I was doing.

"Man," he said, "I never saw a brother in here before trying out for a chef gig."

"Well, you're seeing one now."

He laughed, telling me, "I like your style."

I moved with speed and confidence, though I tried not to overtalk my food to avoid seeming arrogant or intimidating. Even as I prepped, I watched what the other chefs were doing, and I wasn't impressed.

Two hours later, I began putting up dish after dish for the executive chef and his sous-chef. They ate everything. In the end, Chef Capone told me he'd enjoyed my food and would get back to me in a couple of days. He never did.

After two days went by, I called Grant McPherson's Bellagio office in an attempt to secure an interview. Bellagio was the premier resort on the strip with the best restaurants with some of America's most famous chefs. Also, it had a unique feel that set it apart from every other resort in Vegas.

McPherson's secretary had received my resume and set me up to meet with Grant's executive sous-chef, Kevin Thompson. That didn't give me any confidence—I had never interviewed with a number two man before.

Kevin Thompson was self-taught and had come up through the L.A. ranks like I did. The guy I talked to at the Bellagio's employee entrance on my first day in Vegas said Chef Thompson was a hard interviewer with an intimidating reputation, but I tried to remain optimistic. All the same, I was nervous as hell. I went to the bathroom and paced the floor, having self-talk to get the knots out of my stomach.

Chef Thompson was six feet four and had blond hair and blue eyes. He shook my hand firmly and looked me in the eyes. His office was directly across from Grant McPherson's, and I could see all sorts of plaques and medals on Grant's walls. My stomach knotted again. I was on my own. No Gadsby, no Sterling. I'd gotten a lot of jobs on the hookup, but I was a stranger here.

Right off the bat, Chef Thompson hit me up with a blizzard of culinary questions: Who are the top chefs in America? What is your style of cuisine? What was the last cookbook you read? I'd been through it many times, so I knew the answers. I told him about my career path, explaining that I understood that my culinary skills hadn't been fully developed, but that I was still a strong cook and leader.

I was expecting him to ask me to cook for him, like most chefs, but that never happened. Instead, he asked me to name the five mother sauces.

"Hollandaise," I began confidently, "espagnole, béchamel, velo—vel— v—" I began to stammer, trying to pronounce *velouté*. Mario had taught me the five mother sauces and many others besides them, but I'd rarely had to pronounce the French names correctly. As soon as velouté started to give me trouble, I became so unnerved that I completely forgot what the fifth sauce was: allemande, which is just a velouté thickened with egg.

Kneading my hands together and breaking out in a light sweat, I said, "Chef, I can't think of the fifth one right now, but I can make all of them."

I went into a hard sell after that, but I knew I had blown it. I felt disappointed but I maintained my composure. As we wrapped the interview, I assured Chef Thompson that I was a leader and a confident chef in the kitchen, who commanded respect from kitchen workers and chefs all over Los Angeles.

He didn't buy it. He asked how long I would be in town and said he'd call me. Of course, he never did. Even worse, no one called me.

The next morning I prepared to check out of my room and return to L.A. I avoided calling Stacy because I just felt too devastated. My career had just slammed headfirst into a wall and it was my own fault. My anger had taken me from manning the stoves in some of the most prestigious kitchens in Los Angeles to getting rejected by every top hotel on the Las Vegas strip. My bags were packed when a call came through from the Jockey Club's front desk with a message from Caesars Palace: They received my resume and wanted to interview me.

As excited as I was at this opportunity, it was funny to think of working at Caesars after all the high rolling I'd done there in my player days. I notified my probation officer that I wished to stay an extra couple of days for the interview and set up a 9:00 A.M. appointment for the next morning. That night I did my whole recon thing and psyched myself up for the meeting.

As I entered the hotel, I realized how much Caesars had changed over the last fifteen years. They'd added a new tower and the whole place had a very corporate vibe to it. A lot of the shady characters that used to populate the place were missing. Even the bank of safety deposit boxes in the center of the main casino—where I'd once kept my jewelry, cocaine, and cash—were gone. What struck me most was that there were kids running around everywhere you looked. Back in my day, Caesar's was *not* family-friendly.

My mind immediately shifted back fifteen years—to Carmen and me and the Twins pissing away thousands upon thousands of dollars like fools; to the sexcapades we had in the high-roller suites. And here I was back, interviewing for a job.

I knew as soon I entered the executive chef's office that he was unlike any corporate chef I had ever met. He had a rough yet easy way about him. I felt comfortable giving him the real me. We really vibed. He asked me to do a tasting right off the bat.

After the biggest tasting of my life was complete, and I was offered a position at the world's most legendary casino resort, some of my former confidence returned. Though Caesars wasn't the Bellagio, it was right next door. I never took my eyes off the Bellagio, but Caesars would be the birthplace of my Las Vegas career. They offered $40,000 a year and started me at Terrazza. I thought, *God has blessed me once again.*

I let Stacy know that I'd landed the position and would be starting in a week. She was happy as hell, although we couldn't move our family to Vegas until I had collected a few paychecks. Even more good news came when my probation officer informed me that, based on my job offer, a district judge had agreed to terminate my probation two years early.

Yes! No more drug tests, no more searches, no more questioning of my private life. I'd finally been cut free of Big Brother's chains. Though I'd been out of prison for three years, my probation was always a constant reminder that the slightest fuckup would put me back in the hands of the Feds. They'd been my overseer, but that supervision had also kept me straight and focused, and I wanted to prove to the world that I was a redeemed man who deserved his freedom.

That last week in L.A., free at last, I began to wonder what had become of some of my old friends. From time to time I'd hear from someone, but I'd always kept my distance. Now, I was no longer prohibited from contacting them and, more important, there was no danger whatsoever that they could tempt me to street hustle again, regardless of what they were up to.

I had lost touch with everyone from my crew, but managed to track down my old friend Hump down in San Diego. We cut it up. We talked about the past, and he told me how the game had changed for the worse. I had lost contact with T since he took me on the shopping spree when I first got home, and I wanted to see him. Hump put me in touch with T that same day. I made my way back to L.A. to have dinner with T at his favorite joint, Roscoe's Chicken and Waffles in Hollywood. We ate, laughed, reminisced, and talked about the future. T had changed. Besides that he was starting to really show his age, his mind was different. He was subdued, a little reserved, not like the old T at all. He wasn't trying to control every moment like he used to, and he even called me Jeff instead of Hard Head. He was still a hustler at heart, but he wanted to be fully legit. And I knew he was trying.

After that day, I'd see T from time to time, sometimes even when I least expected it. I would be rushing out of my restaurant and T would just be there waiting for me. At first it caught me off guard, but it was cool. It was like my big brother was back watching over me. Eventually I figured out that he was in town because his aunt worked at a local Macy's, but all the same it was good to see him when he showed up.

After that last week at home with my family, I packed up my truck and made my way to Las Vegas. I stayed with an old buddy of mine named Levi Jones. Levi had found his own redemption during twelve years in prison on drug charges, and he was kind enough to give me the use of his sofa and bathroom.

The next three months at Terrazza were chaotic. As a black man walking into an Italian gourmet room . . . you can imagine. I didn't have a strong Italian background, but I didn't have any problem saying, "You know, Chef, I've never done this before."

The head chef was Alberto, this Colombian guy who thought he was Italian. He started me out with some tough challenges, like cleaning a 400 pan of John Dory fish—which is a lot of damned fish—and making fresh mozzarella cheese. My biggest challenge, though, was getting all the cooks in line.

Their ringleader was this fat Mexican dude named Hector, who didn't want to cooperate. Every damn time I looked his way he was sizing me up. Hector and all his boys were always drinking alcohol on the line, and they were always trying to undermine me.

One day one of them offered me a drink in the kitchen. I told him, "I don't drink." After that they all started sneaking it in coffee cups. They were trying to get hi-tech on me like I'm some punk. I went to Chef Alberto about it but he didn't want to get involved.

I couldn't figure what it was, but they seemed to have something on Alberto and he wouldn't back me.

The only way to get Hector's gang under control was to use a little gangsterism and intimidation myself. I waited until the next day that Alberto had the lunch service off while I ran the crew by myself to make my move.

I'm in the back prepping and I see someone bring "coffee" to Hector and some of the boys. I follow Hector into the walk-in freezer, and step right up to him. I can smell the alcohol on his breath. "What's up with you?" I ask him. He doesn't respond.

"There are three ways we can handle this," I tell him. "I can write you up. I can bring you to human resources. Or we can take this personal problem of yours—we both get off at eleven—and we can handle this. Don't let this face fool you; I will fuck you up."

As we squared off in that freezer, my adrenaline was pumping the way it used to back when I was running from the police. But I wasn't a criminal anymore, and there was no one I had to run from. I let myself chill, let the situation diffuse for the moment. I wasn't backing down, but I walked away.

That night when Alberto comes in, I take him to his office, saying, "This is bullshit. Hector's fucking intoxicated right now." Then I'm grabbing the phone, telling Alberto, "Fuck this. I'm ready to report him for drinking on the job right now."

Alberto grabs my wrist. He says, "Chef, don't do it!"

The man pretty much begged me. He was making about $80,000 a year and he wasn't trying to fuck that up. I let it go. But, again, just for the moment.

A couple days later, I'm opening up for lunch and I see a drawing on the blackboard where we listed the daily specials for the waitstaff. It's a rendition of a black man: big lips, Afro, with glasses. Looked just like me with hair.

I called security and employee relations—right after I snapped a picture of the caricature with the digital camera I always keep in my knife kit.

Over in employee relations, this woman asked me if I felt offended. Of course she wanted me to say no so that I couldn't sue.

"I look at it in different aspects," I said. "One aspect, it's a picture of a beautiful black man. The other aspect is that it's a racist gesture and a ploy to get me to bitch up and ask to be moved to another restaurant."

Telling her I didn't want a transfer, I returned to the restaurant, where I spent the next week putting the fear of God into those motherfuckers—watching their every move and getting in their faces if it looked like they were even *thinking* about stepping out of line. After that, the executive chef called me in for a meeting.

At first, I was afraid someone had complained about the strong-arm tactics I'd been raining down on Hector and his crew. But that didn't seem likely, since I was always careful not to do it in front of witnesses.

"Jeff," he said. "What do you think of the Paladian Buffet?"

I said, "I don't know. I never worked at a buffet before."

"Well, you're going to run it," he said. "I'm promoting you to chef de cuisine there."

I didn't know if he was trying to make sure I didn't sue the hotel, if he was protecting his friend Alberto or making a PR move by making me the first African American chef de cuisine at Caesars Palace. Maybe it was a combination of all three. All I knew was that I'd played the corporate game and come out on top. I was going to be the first black man to run a restaurant at Caesars Palace in its thirty-eight-year history. And my ninety-day probation wasn't even up yet.

At $50,000 a year, it was the most money I'd ever made legitimately. The first thing I did was to call everyone who'd been supportive of me since leaving prison and express to them my gratitude. I called Stacy, then my mom and dad, and Robert and Sterling. I couldn't stop talking; I was riding high.

After another month, I saved enough money to move my family into a brand-new three-bedroom, two-and-a-half-bath house in the Las Vegas suburb of Henderson (which seemed like good luck since it was my name).

The Sunday night before my first day on the new job, I stood in line at the Palladian Buffet like a regular customer to check the place out. It seemed a bit unorganized and hectic from the front of the house, but that was nothing compared to what I discovered behind the scenes in the kitchen the next day.

I met with the sous-chefs who would be working under me on Monday morning to be brought up to date on the Palladian's standard operating procedures. The first thing I learned was that the Palladian *had* no formal operating procedures. Likewise, there was no useful information whatsoever on the kitchen's computer—no set menus, no job descriptions, no updated ordering forms, not even a readable inventory document. Whether it was sabotage or just some of the worst management I'd ever seen, I knew I was going to have to clean house.

The staff was a melting pot of senior citizens, middle-aged kitchen vets, and young, unmotivated wannabe gangsta types, fifty-five in all. The average employee had fifteen years with the company and had grown far beyond complacent. They were stealing, slacking off, and screwing up the food.

The first thing I did was investigate the kitchen's politics— find out who the leaders were in each area, sit down with the union shop steward, and determine who I could extract information from. And it had to happen quickly if I was going to avoid showing any signs of failure.

Within hours, I'd built an alliance with several members of the crew. One by one, they began to stream into my office, a lot of them only too happy to inform on their coworkers, telling me who did what and how they did it. I didn't trust that type. To me, it wasn't the message but the messenger. I wanted information, but I wondered what motives they had for telling on their coworkers without even being asked to do it.

I chose to roll the dice with a couple of employees by bringing them into my confidence. One, who was from the West Side of Vegas, which is the hood, was very street and had kept her eye on me since the first moment I stepped in the kitchen. I knew right off I could have her on my team. The guy, a Caesars vet, was very quiet and seemed to have the most control of the breakfast and lunch shift. He had respected me right off. I broke the rules of management and met each of them outside of work and asked them for the rundown. We went over the schedule and they let me know about each and every Palladian employee. They told me about workers stealing lobsters and selling them to the dealers as club sandwiches in the casino; cooks and waitstaff trading food to the ladies in the dry-cleaning department in exchange for dry-cleaning their personal clothing; the theft of steaks and shrimp; and how some employees would swipe time cards for people who had stayed home sick. Basically, the entire kitchen ran the whole nine yards of culinary corruption.

I was taking a risk by meeting with these two, since I had yet to establish a trust with them, but it was worth the gamble. This was the first chance I ever had to run an operation on my own, so failure was not an option. Within forty-eight hours I started making drastic changes and establishing my own standard operating procedures by adapting the ones I learned in prison and in L.A. to the Palladian. I wrote new rotating buffet menus and began retraining my cooks. I even put out several prison entrees, such as Friendly's Buttermilk Fried Chicken, the dirty rice from Terminal Island, and even the shit-on-the-shingle. To earn the respect of my team and have them follow my lead, I came in every day with my knives and other tools and jumped right on the line with them. I showed them I could cook, as well as motivate, and that I was willing to work hard to make the buffet a respectable eatery, as it once was in my high-roller days.

The kitchen staff had developed some very bad habits over the years, but some of them were still teachable. I showed them proper grilling and seasoning techniques, roasting times, and other secrets of batch cooking that I'd learned along the way.

The staff were so unmotivated and untrained that they were cooking food three and four hours in advance and putting it in warmer boxes (used to keep food warm until it is served, but only for a short time to avoid bacteria), and sometimes putting raw meat into the oven without seasoning. Many of the cooks were lazy, poorly groomed, and unchallenged. There were few, if any, staff meetings, and no hierarchy had been established. The previous chef used so much frozen food that working with fresh ingredients was unfamiliar to them. There were nine walk-in refrigerators and two walk-in freezers—each the size of a one-car garage filled with pallets of food—and not much space for fresh foods, which meant that not much was being used.

The pantry was run by six or seven women who all hated and snitched on one another. They were a pain in the ass, but I had to be careful about how I dealt with them because they were notorious for setting up chefs. Remembering how I let my anger and immaturity mess up my career at L'Ermitage, I never met any of my female employees alone in my office or even spoke to them without a witness present.

Still, nobody screwed around with me. They'd heard about my battles at Terrazza and knew I was a no-nonsense chef who knew the game. They knew change was coming.

As the months went by, the buffet improved noticeably. The morale of the employees, and the food and presentations improved at the Palladian. I instituted fresh herb baskets on the line and started a batch-cooking program; that way everything was cooked in smaller batches and so the food spent little time in the warmer boxes. I made sure that no meat went into an oven before a chef saw that it had been properly seasoned.

About four months after I took over, a twenty-year-old brother named Lamont came to me during dinner service and said, "Chef, can I get an early out tonight? I want to do something special for my birthday." He was a food runner who had an aunt who worked in the hotel.

"All right," I told him. "That's not a problem, but first make sure that everyone's had their break and that all the stations are covered. After that, come see me and we'll see about getting you an early out."

"No problem, Chef."

Three hours later a couple of the other runners came to my office to tell me that Lamont was nowhere to be found and showed me his walkie-talkie. He had left it in the dining room and gone home without checking with me. I suspended him immediately and got the ball rolling toward his termination. A part of me felt bad for a young black man trying to make it. But I knew that at the end of the day, if I didn't do my job, I'd be the one getting fired.

After that I became a marked man. Most of the African American staff in the food and beverage division stopped talking to me. Everyone looked at me like I was an enemy to the black race because I fired a brother, but, hell, there wasn't anybody watching *my* back or taking the heat for me. I didn't care anymore if they liked me or not. I still worked with them, still looked them in the eye when I walked by, and I still treated them with dignity and respect, but it would be another year before they started returning that respect. You'd think they would have been proud to be working with a black chef—I wanted them to be proud of me—but some of them were like crabs in a barrel.

One brother even got in my face, saying, "Who do you think you are coming up in here and giving us hell?"

A couple years back, I probably would have lost my shit and knocked him out. But now that I was in charge, I had to stay cool and try to identify with the people working beneath me. I calmly told him, "Listen. Don't hate on me, hate on yourself. You've been here more than twenty years—you should have *been* the chef by now."

A year and half went by and the public had taken notice of all the improvements at the Palladian. I had improved the food and the look of the place (even added new plants to the dining room). There were many new items that the customers loved, like gumbo and braised beef short ribs, and business was great. The American Tasting Institute in San Francisco contacted me to invite me to an award ceremony. A number of Las Vegas chefs, including myself, were being given some kind of medal of excellence. I was told to make sure I'd be there.

It was a big deal to me, since I'd never been honored in any way before. I had Stacy get dressed up in her best outfit and I put on a brand-new Caesars Palace chef jacket, and we made our way to the ceremony at the MGM Grand.

After the organizers had called out all the medal winners for culinary excellence, they announced that they would also be naming the best chefs of the year 2001.

Next thing I know, they're saying, "Jeffrey Henderson" and everyone is applauding and Stacy is hugging the hell out of me. For a moment, I wasn't even sure what was happening. I had just been named Las Vegas Buffet Chef of the Year.

I was in awe. I thought, *This can't be right,* as I walked up to the dais. It was an incredible moment for me; but even as I was receiving that second medal, I was thinking about getting another award—and the next one wouldn't be for a buffet.

The most important thing about my award was that the exposure it brought gave me opportunities to work raising awareness in the community. The time had come for me to start giving back as much as I could to society, as I'd tried to do in prison with the teenage awareness program. I began to build alliances with other chefs and nonprofit groups around Las Vegas. I was introduced to members of the city council and started appearing on talk radio and local news shows to discuss the plight of black youths as well as ex-offenders. The Nevada Gaming Commission even invited me to address its members regarding the problems facing ex-cons working in the industry. I made appointments with school district officials to set up events and speak to their students about making the right choices and decisions.

This led to my founding my own nonprofit organization, the West Side Foundation. At every school and prison that would have me, I told my personal story of redemption. Although I felt I did a great deal to redeem myself personally, I also knew that I had an obligation to do my part toward stemming the tide of hopelessness and criminality that had plagued my community for so long. Because of the influence my generation had had on America's youth by initiating the crack epidemic, I knew I still had a very steep debt to pay off.

I wanted to use my story, my work, and my success to inspire and motivate at-risk youth. My mission was to help individuals who society had given up on, who the odds were stacked against, to get a second chance. After I spoke to students at several schools across Clark County, the media took notice. The local press and TV stations began running stories about my life, my cooking, and my work with the community. Before long, I was doing as much public speaking as I was cooking.

Three years into the job, Stacy was pregnant with our third child. In March 2002, Troy Kennedy Henderson was born. She looked just like her mother. We were doing well financially and our lives were great, but I still wanted to further my career. I also knew that I had been cheating my family out of my time. With the demands of my job and community work, they only got to spend a few days a month with me. Stacy started putting pressure on me to spend more time with them. I entirely agreed with her that I needed (and wanted) to, and ever since then I've done all I could to try to balance my time spent with my family and career.

I had been telling the executive chef for some time that I wanted to run Nero's, a fine dining steakhouse and the top-grossing restaurant at Caesars, and he constantly assured me that it would happen. In anticipation, I studied the menus and learned the entire operation inside and out. Finally, the executive chef told me that the chef there was leaving and that I should get ready to take over, but almost as soon as he offered me the job, circumstances got in the way.

While I was remaking the Palladian, Chef Kevin Thompson left the Bellagio to run Caesars Café Lago. And just as I was preparing to take over Nero's, Kevin quit and took a job in L.A. When the executive chef asked me to fill in temporarily for Kevin at Lago, an all-night café, I knew that I would never become chef at Nero's. As much as the executive chef admired me, he was going to sacrifice me to keep Lago up and running. The daily grind at Lago quickly burned me out. I was spending fourteen hours a day dealing with scoundrels and cooks with no passion. For the four months that I ran Lago, there were only a handful of good cooks around me. It was frustrating.

Finally, I told the executive chef that I was ready to move on and accepted a position at the Hard Rock Hotel as executive chef of A.J.'s Steakhouse. It was my first gourmet room in Vegas, but I grew bored with it within months. I was never allowed to change the menu, and doing the same thing every night made me feel like a robot. On my off days, I went to L.A. to cater private parties for Hollywood's black elite. Clients included Terry Lewis, Dorian Wilson, Teshia Campbell, Larkin Arnold, Aftermath Records, and the Tavis Smiley Foundation. Working for myself gave me the opportunity to continue to refine my cuisine—sometimes I made good money doing it, but mostly I found that I was drawn to events that didn't pay much but were for a good cause: helping people or furthering my career. My nights at the Hard Rock just felt like drudgery and busywork by comparison.

One day, I had had it. I needed a day or two away from the grind at the Hard Rock, so I went to see the executive chef of the Hard Rock Hotel, saying I was sick. He wasn't buying it. I had never called in sick anywhere, but I was just tired of it. When I came in the next day, the executive chef tried to fire me, saying I hadn't called in. It was bullshit. I had come in personally to tell him I wouldn't be able to work that day. We agreed on a severance deal.

After my severance ran out, I decided to hit L.A. to cater private parties full-time. I started my own company in Los Angeles called Posh Urban Cuisine, though I was still living in Vegas; but that didn't last long because I could never get people to pay their bills. People wanted to get everything for free. When I asked for my money, they gave me some nonsense about how it was worth my time just for the "exposure" they were giving me. They were some of the richest African Americans I'd ever met, as well as some of the cheapest people on earth.

I sucked it up and accepted yet another hotel position . . . at the Holiday Inn in Brentwood, California.

I never imagined in a million years that at this stage in my career I'd be working at such a run-of-the-mill hotel, but they were paying me $55,000 a year and I was executive chef at a beautiful rooftop restaurant called West. The other good thing was that it was a low-stress position that didn't require a lot of my time.

The staff at West all had good hearts, but no one knew how to cook. Everything came from Sysco, the largest and most commercial food purveyor in the country, all of it frozen. The only thing that seemed fresh was the water. It was the first time I was ever ashamed of working somewhere and I never told anyone I worked there. People would ask where I was and I'd say, "West Restaurant," without telling them where it was.

After a quick four months I made my way back to Vegas. My timing was right because word was Grant McPherson had resigned from the Bellagio to open Wynn Resorts. With both him and Kevin Thompson gone, I might have a better chance with the new food and beverage team looking for new blood.

Wolfgang von Wieser from the Four Seasons Las Vegas was the new executive chef of the Bellagio. I met up with a friend who was the Bellagio's banquet chef, and he hooked me up with a meeting with Wolfgang and the VP of food and beverage. They immediately asked me to do a tasting.

I hit the Café Bellagio kitchen a day in advance and set up prep right next to the other cooks who were working. I wanted to make a statement right off. I wanted to show the crew how I rolled. The

Café chef was off that day, so I mingled with the front and back of the house in an attempt to get the rundown on the kitchen politics. After talking with the cooks and waitstaff, I was convinced that they didn't have much confidence in the chef. I pulled out my six-course prep list and started my meticulous mise en place.

After reducing my sauces and braising my short ribs, I wrapped my first day of prep. I secured my cart and placed it in the walk-in.

I arrived promptly at 9:00 in the morning, even though show time wasn't until 4:30 in the afternoon. I searched out the best china the Bellagio had to offer. After gathering imported china from several of the gourmet restaurants, I was ready to cook.

When Wolfgang, his executive sous-chef, the vice president of food and beverage, and his director rolled in the kitchen, all eyes were on me. They were all standing at a table right in front of the hot line, so they could see everything. I was onstage, but couldn't have been more confident. My station was tight, everything was in place—as Robert always told me, your workstation and appearance are a reflection of your professionalism. I made sure that from my starched white Brigard chef coat to my clean shaven face and polished clogs, I looked my best.

I hit them first with my signature amuse: a seedless watermelon cube, marinated in Remy red cognac, and topped off with a minted citrus salad with a streak of reduced aged balsamic syrup. Then I put up my first course, a sushi-grade Ahi tuna terrine with avocado fondue, crispy leeks, and a warm ginger-soy broth.

The second course was a roasted quail soup. Chef Wolfgang had asked me to make a chicken noodle soup, but I didn't want to do anything that simple, so I decided on something with a twist. I served the quail boneless and in its natural juices with wilted greens and pasta.

My third course was pan-seared striped bass, fingerling potatoes, and garlic spinach with tomato water, garnished with microgreens and served in a bone white bowl. Halfway through the tasting, the brass appeared to be enjoying my food. I watched them talk among themselves as I plated course after course.

Next up was maple-braised beef short ribs, served "off the bone," with potato puree, Swiss chard, roasted cipollini onion, and finished off with reduced braising juices.

Though it wasn't required, I hit them with a grand finale—a dessert of chocolate banana bread pudding, vanilla ice cream, and warm caramel sauce served with a cognac milk shake.

The trial was over, and the jury huddled for a verdict. Their deliberations lasted less than a minute. I had a job at the most prestigious resort ever built on the strip. They named me the executive sous-chef at Café Bellagio, a twenty-four-hour operation that was doing more than $20 million a year in revenue.

It was time for a real celebration. Stacy, myself, our three kids, and my sister-in-law Sugar got dressed up and splurged on a great dinner at my new place of employment. I mixed the family celebration with another recon mission. The Café Bellagio was gorgeous and huge—five-hundred-plus seats—and the whole place bustling with action. Once we were seated, I began to observe the staff and time the food coming out.

The menu was mostly American fare with some Italian influences. I ordered a Cobb salad, Stacy and the kids, being vegans, had steamed vegetables. The food was excellent and abundant. That's Vegas—big food.

When I arrived to start working the breakfast shift the next month, my mind was already overflowing with ideas for new menu items and changes I'd like to make, but I was only the second man in command. Number one was a young Frenchman who was burned out. Vegas had laid its wrath upon him. He had no experience dealing with high volume and I realized that it was only a matter of time before he cooked his last meal there. For all that, the operation was very well organized and the staff ran on cruise control. They were open to me from the beginning.

I wasn't a polished breakfast cook, but even though I heard that the Café banged out thirty-five hundred covers a day in total, I wasn't too worried. The first station I worked on was the fry and grill station, which was a bit of a challenge because the grill was so small, but I mastered it soon enough. The executive chef was always stressed and frustrated, and somehow that took a lot of pressure off me because I knew I could do my job—and his. The poor guy was very emotional and soon started confiding all of his troubles to me. If I'd had any doubts that he didn't intend to stay and tough it out, he personally relieved me of them.

Wolfgang called me into his office less than a month later, asking me how it was going.

"Everything's fine, Chef."

"Are you learning everything?"

"Yes," I told him. "I think I've got it down to a science."

And then he said: "Just be ready."

Days later, I was notified that the chef was leaving and that I would be taking over the reins. A couple of people made bets I wouldn't last long, but to me it was one of the easiest kitchens I'd ever worked at because the staff consisted of a talented crew of Latinos led by Raul, Chico, and Cholo, and two white boys who brought their A-game to work every day. The previous chef wasn't coming back, and his leaving had nothing to do with me, so I didn't have to fight to win over anyone who was loyal to him.

The transition of power was smooth. I revamped a few things, introduced new specials—like mixed green salad with walnuts,

Granny Smith apples, and a balsamic vinaigrette; Maryland crab cakes with shaved English cucumber salad with curry sauce and chive oil; and even my jailhouse Friendly's Fried Chicken—and reworked the menu to ease the burden on the cooks by making the dishes simpler to prepare. But there was no need to clean house or establish myself like a gangster, the way I had in all the other kitchens. No one resented me; in fact, they were all very supportive.

Overseeing the Café, to this day, is a very challenging task, simply because the Bellagio has such high, exacting standards, but it is free of the kind of drama and chaos I'd known in so many other places through the course of my life. And so, at last, is this hard head of mine.

AFTERWORD

GOD BLESS THE DEAD

News Release
Monday, February 13, 2006

1 MAN SHOT AND KILLED,
FIVE OTHERS INJURED

LOS ANGELES: Los Angeles Police Department detectives are investigating the shooting death of Thess Good, a 45-year-old Black man.

On Saturday evening, February 11, 2006, at about 6:55 p.m., Thess Good along [*sic*] and a group of men were in a garage that served as a barber shop to the rear of 10350 Hobart Boulevard. A white vehicle drove up to the location and stopped in front of the location. Moments later, a red Jeep Cherokee drove up to the location and began to fire multiple shots into the garage.

Thess Good and five injured men were taken to local hospitals. Good died shortly after.

The motive for the shooting appears to be gang related. The suspects remain outstanding.

Anyone with information is asked to call 77th Street Homicide Detectives at 213-485-1385. On weekends and during off-hours, call the 24-hour toll free number at 1–877–LAWFULL (529–3855).

To his friends and family, Thess Good was better known as T-Row.

I was cruising home from a double shift at the Bellagio when my cell phone rang. I didn't recognize the caller's number.

Flipping it open, I said, "Chef Jeff," and a strangely familiar voice said, "What up, Hard Head?"

I was taken aback a moment, thinking, *Who is this calling me Hard Head?*

"It's Hump, man."

"Hump!" I said, "Damn, is that you? How've you been, man?"

Hump and his brother Jake were former Skyline Bloods who had been two of the top players on my crew back in my criminal days. It made me nervous to be hearing from him out of the blue, since we hadn't spoken in some time. I wondered how he'd even gotten my cell phone number, but I played it cool and asked him what he'd been up to.

"Homie," he said, "you didn't hear?"

"Hear what, Hump?" Just then I could feel my heartbeat racing through my chest. Because what were the odds that the news from my former life was going to be anything but bad?

"You won't even believe this. T got killed, Homie. Some motherfuckers shot him."

"What?" My voice was shaking. "But I just talked to T two weeks ago—we were going to get our daughters together to play. What the fuck, Hump? What the hell happened?"

"Some fools did a drive-by on a house on Hobart Street where T was getting a cut. Like five dudes and some kids got hit."

"Damn!" I said. "Damn! How could T let someone creep up on him like that? T's too sharp to get caught up in crossfire. Hump, shit, maybe you got the story wrong."

"I don't think so, man."

When I hung up with him, I started working the phone to try and find out what went down, still holding on to the hope that it was just a false rumor.

After getting nowhere on the phone, I went to track down T's aunt who worked at a Macy's here in Vegas.

As I walked into the store, I saw her in the distance adjusting some suits on a rack. When I got closer, I knew the worst was true—I saw T's death in her face. This lady always had a radiant smile, but now her features were tight and pained.

Her first words to me were about how much her nephew had favored me.

"Everyone always thought you and Thess were blood brothers."

"Yes," I said to her. "It's true."

Her eyes filled and she began to weep. We hugged as we talked about how much T had changed his life and found God, about how he had been working as a longshoreman in San Pedro, right next to the prison where I had changed my life. His children had become the force behind his deep desire to kick the street jones, the hustler's life. T had even gotten married. Like me, T had been ready for the second half of his life.

But T's soul kept him connected to the streets. His love for the homies ran deep in his veins. If only he could have loved the homies and the streets from a safe distance, his life just might have been spared.

Parting ways with T's aunt, I was crushed, devastated, and my mind raced back in time—to all the fun we'd had, all the wrong shit we had done, and all the times death had come knocking on our doors. Though T and I hadn't hung out much in the last eighteen years, I still loved him as I did when I was a mixed-up youngster looking for a role model and a quick avenue out of the hood.

Driving home that night, I felt a deep, dark need to seek revenge for my brother, to track down the killers who had cowardly shot him down in cold blood. But I maneuvered my way through the hate and the anger. What would it prove? Besides, killing was never in me, and that world was no longer mine. My love affair with the streets was another dead thing. Where I once saw romance and profit, now I could only see madness.

Five days after T-Row's murder, I began the four-hour journey across the desert to South Central L.A. to pay my respects to T and his family.

My wife and family were dead set against me making the trip; they knew that violence was always a distinct possibility at a hustler's funeral, even if he had retired from the game. But I had to go.

It was raining that morning as I approached the church. Cars had begun to line up all around it. I circled the block just to check the surroundings in case I had to make a quick getaway. The sidewalks were teaming with buff brothers of all ages dressed in shiny player suits and Al Capone–style bowler hats. There were also plenty of young knuckleheads who showed up in their street clothes, the blues and reds of the Crips and the Bloods prominently displayed.

The funeral was a family reunion of sorts. Within twenty minutes, a dozen different people had come up to me saying things like, "Hard Head! Hard Head! You remember me, nigga? I hear you doing big things these days for Hollywood and all that."

Their faces rang a bell but I couldn't quite remember who all these old acquaintances were—and I just didn't feel like rekindling a past that I wanted behind me. I just wanted to bury my big brother.

The San Diego crew arrived shortly after I did: Hump, Jake, Silky Slim, and a few more of the old homies. We all grabbed seats together as the church began to fill. By the time the first preacher addressed the assembled, every last pew was packed and the crowds filled every aisle out of the doors and down into the streets. T was on his way to heaven with a rock star's farewell.

I do believe T went to heaven; he loved the Lord and feared him as well.

Flanked by the old Diego crew, I scanned the room and spotted Carmen. She was dressed in black, still looking good as ever. Then I saw the remaining half of the Twins, and thought about the other one who had lost his life to the game that wounded, killed, and imprisoned so many—the innocent and the guilty alike.

Preacher after preacher talked about the killings, the plague of black-on-black crime that had to be stopped. Then a screen was set up and a video montage highlighted T's days of lowriding and living life as we'd once known it.

I felt T's spirit calling to me to speak to his family, to testify. As I approached the podium at the front of the church, I paused to look at him in his casket. He was as clean as always, suited up with a smile, but it wasn't him. I mean, it wasn't *supposed* to be him. He had always been so smooth, had escaped death and long prison stints so many times. But no one was untouchable. I realized then, staring into T's pale, frozen face, that it could have been me in that box many times over. I'd been stabbed, I could have been snatched off the street and tortured to death like one of the Twins, or murdered in prison. It was an overwhelming moment.

I got my head together and told the congregation that T had raised me, instilled confidence in me, taught me how to feed my family and escape poverty, to be streetwise and tough. Although we had done things we weren't proud of, I said, we both made substantial changes in our lives. I talked about all the lives T touched in so many different ways. I told them I would never forget how T saved my life when I went to prison, by reaching in from the streets to put a shield of protection around me.

Gazing out at the crowd, I also wanted to lash out. I felt that someone among them had to know who murdered T, and they had the nerve to come to his funeral and keep their mouths shut. But I held back my emotions and my pain.

Some of these fools are going to learn from this, I thought as I made my way down the aisle. T wouldn't die for nothing. He'd be an example that would convince one of these kids that there was no good end to being a hustler; sooner or later it would catch up with you.

I had barely set a foot outside the church when I realized that I was wrong: Chaos had broken out. A bunch of young knuckleheads snuck out of the service early to get drunk across the street and they were suddenly throwing down on each other. Someone flashed a gun and the L.A. County Gang Units rolled in, in force.

Carmen came up next to me as I stood just outside the church and we hugged. It was a weird feeling seeing her. I didn't really know what to say. My mind was racing; I just wanted out, away from my past. I told her I'd call her soon, and then I quickly made my way to the car and fled the area before it got even more out of hand.

They couldn't even let T rest in peace on his final journey home. Those fools left me feeling disgusted and greatly depressed, like I had witnessed the beginning of yet another cycle in the rarely changing fate of my community. I didn't even go to the cemetery to see T buried.

Silky, Hump, some of the other old homies, and me paid our last respects over brunch at T's favorite restaurant, Roscoe's. Reminiscing about the old days over waffles and fried chicken, though, I quickly found myself drifting out of the conversation. I just didn't fit in anymore. Moreover, I began to feel uncomfortable. The conversation felt like ancient history, and I had moved forward. I just wanted to let it go once and for all.

I had barely touched my food when I stood up, dropped some bills on the table, and wished them well. I didn't want to talk about the streets anymore, didn't want to hear about the street anymore. And while I didn't want to come off like some preacher man or rub my success in their faces, I just wished they all had bigger and better things on their minds. But it's hard to tell a man in his forties who's been hustling all his life to change. And it's not something that can just be done overnight. Hell, it took me eighteen years.

I just wanted to get on the highway and get back to my family and my restaurant—back to my life.

Day turned to night as I crossed the desert. I thought about all the programs I take part in that try to effect change in youths before they find themselves too deeply trapped in the cycle of crime and violence. I was taking the next day off to interview some potential employees and work on a youth presentation I had coming up. I thought I could use T's death to slam home the consequences of thug life.

"I just buried my longtime friend," I would tell the youngsters who wanted to emulate the hustlers that T and I had once been. "If we can change, you can change."

Once I got back to my kitchen that Tuesday, I reflected on how blessed I was to live and work with normal, everyday folk. I don't have any homies anymore. I have friends and colleagues, and I have my family.

These days, I get up at 4:30 every morning, hit the gym, and get to the kitchen by six. Before morning service gets heavy, I line up my staff, get them psyched, and boost their morale. I check my groceries and prepare my daily specials. Throughout the day, I tour the front of the house and check in with my customers.

I always phone Stacy and the kids at the end of the day to tell them I'm on my way home. When I get out of the truck, the kids are always waiting at the door—Jeffrey Jr., who is eight; my little girl Noel, seven; and my youngest daughter, four-year-old Troy. They show me their drawings and homework and I pick their brains about what went on in school that day. All the things that were missing from my own childhood.

Between the Bellagio and my youth projects, I don't have a lot of free time, but I save Sundays for my family. I load the kids' bikes in the trunk and we drive out to the park or up into the mountains. We also love to hit all the resorts and go swimming in their beautiful pools.

The children all have extracurricular activities of their own. I'm Jeff Jr.'s biggest cheerleader when he plays basketball for the city of Henderson. Twice a week, I take my girls to their ballet classes—Troy wants to be a ballerina. I film my two little girls spinning on their toes and think, *This is what it's all about.*

A central part of our life is dinner every night. Stacy usually starts the cooking and I jump in to put on the finishing touches. The kids get excited whenever I hit the stove. They shout, "Daddy's cooking!" They cheer, "Put on your chef's coat!"

I do and then ask little Troy, "Who's Daddy?" and she says, "Chef Jeff, Bad Boy of Cuisine." It puts the biggest smile on my face.

We eat and talk for a while, and Stacy and I always make sure our kids see how important education is by discussing school with them every evening. Sometimes dinner isn't over for several hours.

Noel is very inquisitive. She'll ask me, "Did you have fun at work, Daddy? What did you cook?"

When they get older, I hope they will remember that Stacy and I spent the time with them. I bang twelve hours a day for them. I lived twenty-three years on the streets. Now I live for my family. And that's the best kind of living there is.

ACKNOWLEDGMENTS

There are so many people who helped shape my life during and after my imprisonment, but I can only acknowledge some of them.

Grandma, you have always been my inspiration since my childhood. You encouraged me to stay strong when I was down and you were there for me through every phase of my journey. Thank you for keeping the cookie jar full for us, and for always accepting the collect calls to hear my cooking stories. Thank you for believing in my dream to become a chef. Your powerful spirit lives within me.

To my mom and dad, thank you both for never giving up on me. I know in my heart that you did your best. To my sisters, thanks for the love and support through it all.

My son Jamar: I apologize for not being there for you—your football games, your graduations, your first date, and the times you needed your father as you were growing up. Every day I think about how I can make my absence up to you, but know that my love for you has never wavered and I will never go away again.

Jeffrey Jr., Noel Marie, and Troy Kennedy: I work endlessly for you. I savor every moment we spend together.

To the brotherhood in prison: Thank you for your inspiration and your direction in making me see the positive in myself.

I appreciate you for introducing me to knowledge that was once so foreign to me. You awakened me.

Big Roy opened my eyes to the prison food world, and Friendly Womack helped me dream that I could really become a chef on the outside.

To all my counselors and caseworkers over the years who believed in me, thank you. A special thank-you to Mr. Hershman, the most passionate educator in the penal system. You really made an impact on me by getting me to look at the man in the mirror. Thanks for your continued support.

Robert Gadsby, Sterling Burpee, Marcus Samuelson, G-Garvin, and Patrick Clarke gave me the courage and inspiration I needed to become a chef. Thanks to all of you for opening the doors for me and other African Americans who want to become chefs, not just cooks. To Robert and Sterling, thank you for having the courage to bet on a thirty-two-year-old ex-con. Robert, you gave me the necessary road map, believed in me, and invested endless nights on the hot line shaping my talents. Sterling, we cooked hard at the LAX Marriott; thanks for bringing me in the back door to prove the critics wrong.

Sara Bowman, Garry Clauson, Josh Thompson, Tom Hanson, Jim Perillo, Sgt. Norma, Mr. Payne, Bradley Odgen, Wolfgang Vonwesier: you have all influenced my career in tremendous ways. Thanks for believing in me.

Thanks to all of the cooks who I banged with on the line. You guys are true soldiers, who pushed my confidence over the top. Mario and Feliciano, you should be James Beard winners. Thanks for getting me ready for Vegas. The food industry wouldn't be anything without guys like you. Keep cooking hard.

Super Agent Mike Psaltis of the Culinary Cooperative/Regal Literary made this book a reality; you pursued having my story told with integrity and passion. Thanks, Mike, for the hard work and the endless e-mails about this project. Ian Spiegelman is the man who helped me shape the book and tell my story in a raw, compelling way that I hope inspires many people to make their dreams become realities.

Thank you Henry Ferris, Lisa Gallagher "Miss Brit," Harriet Bell, Dee Dee DeBartlo, and the whole crew at William Morrow/ HarperCollins. You all really put love behind the story and me.

Stacy, my fabulous wife, you've stayed down with me through it all. Thanks for believing in my dreams and letting me pursue them hard. I love you and our children unconditionally.